P9-CAD-572

About the Author

Photo by David Finch

David Finch loves stories. He is a consulting historian with more than twenty books and articles to his credit. As the author of several books about the Canadian oil industry—including *Hell's Half Acre: Early Days in the Great Alberta Oil Patch* and histories of the Canadian Society of Petroleum Geologists, the Canadian Society of Exploration Geophysicists, and the Canadian Association of Petroleum Landmen—he is always researching the history of the "patch." In 1999 the Petroleum History Society of Calgary awarded Finch the Lifetime Achievement Award for Dedicated Investigation and Professional Scholarship in the Study of Canadian Petroleum History. David loves the out of doors and with his wife, Jeannie, and daughter, Annie, has camped, hiked, skied, mountain-biked, and canoed in many parts of western Canada. David lives with his family in Calgary. ∎

PUMPED
EVERYONE'S GUIDE
TO THE OIL PATCH

David Finch

FIFTH
HOUSE

Cover and interior design by Dean Pickup
Cover illustration by Dean Pickup
Edited by Alister Thomas and Meaghan Craven
Copyedited by Lesley Reynolds
Proofread by Dallas Harrison

The type in this book is set in Sabon and Gil Sans.

The publisher gratefully acknowledges the support of The Canada Council for the Arts and the Department of Canadian Heritage.

We acknowledge the financial support of the Government of Canada through the Book Publishing Industry Development Program (BPIDP) for our publishing activities.

 Canada Council **Conseil des Arts**
for the Arts **du Canada**

Printed in Canada on 100% recycled paper

2007 / 1

First published in the United States in 2008 by
Fitzhenry & Whiteside
311 Washington Street
Brighton, Massachusetts 02135

Library and Archives Canada Cataloguing in Publication

Finch, David, 1956-
    Pumped : everyone's guide to the oil patch / David Finch.

Includes bibliographical references and index.
ISBN 978-1-897252-09-3

    1. Petroleum industry and trade--Canada. 2. Petroleum industry and trade--Canada--History. I. Title.

HD9574.C2F565 2007   338.2'72820971   C2007-903483-7

Fifth House Ltd.
A Fitzhenry & Whiteside Company
1511, 1800-4 St. SW
Calgary, Alberta T2S 2S5

1-800-387-9776
www.fitzhenry.ca

Contents

Foreword   By Andrew Nikiforuk   vii

Hi There   An Introduction to This Book   ix

Jargon   The Roughneck and the Pump Jack   1

Discovery   Oil Is Where You Find It   17

Gushers   Drilling for Oil   33

Consumers   Using Twenty-five Barrels of Oil   45

Conflict   Seven Examples of Canadian National Energy Programs   61

Outrageous   The Price at the Pump   85

Bitumen   The Oil Sands   99

Frontiers   Oilfields on the Edge   121

OOPS!   Mistakes and Lessons   137

Booms   This Time It's Different   163

Glossary and Index   183

Foreword

Canadians have never thought much about energy unless the price at the pump emptied their wallet. That's happening a lot these days. I'm offended, and you, dear reader, are probably getting ticked off, too. But if you read this book, you will understand why more offensive price hikes are on the way. The reason, aside from this thing called "The End of Cheap Oil," is pretty basic: Canadians just don't think strategically about energy and are about to pay a hell of a price for their collective ignorance.

David Finch, a crafty oil patch historian, thinks Canadians should get informed instead of mad. He believes that energy, carbon, and water—the unholy trinity of modern life—has transformed Canada into a so-called energy superpower with some super liabilities as well. The realities are stark. Oil is getting harder to produce. It's also slurping up more and more surface- and groundwater. And it's making hefty clouds of carbon. So, Canadians have to get smart about the black gold they are pumping into their cars, homes, and food.

Finch has got it right: he doesn't think we can really afford to be stupid drawers of water, careless hewers of wood, and greedy oil-sand miners anymore. And that's why I really like this book. No evangelical industry bashing; just the facts. Hey, we're all in this hydrocarbon mess together. So, the amiable historian explains what the energetic oil patch does and why we are all addicted as hell to oil. We're not only the world's largest consumer of petroleum products on a per capita basis but also the globe's largest belchers of climate changing gases. Go Canada!

In addition to providing the greasy details on how the stuff in the ground gets to your fuel tank, Finch gives Canadians a fun guide to petroleum vocabulary (it's sexy), a whole lot of head-scratching statistics, a history of national or not so national energy policies, and some excellent distractions. Did you know for example that Albertans once planned to nuke the entire tar sands with the code-name Project Cauldron? Or that Hell's Half Acre isn't located below ground but on the surface in Alberta? Or that it takes three times as many oil and gas wells to keep the nation running then it did seven years ago? Or that the average Canadian consumes 25 barrels of oil a year? No kidding. An ordinary family of four has a 100-barrel-per-year addiction. And those silly Albertans, well get this, on average, each one of them burns his or her way through 60 barrels a year.

So read on. Get pumped. And power down.

Andrew Nikiforuk, Calgary, May 2007

Hi There

An Introduction to This Book

Do you find the oil industry confusing and hard to understand? Me too.

Petroleum is oil. It is also gas, heavy oil, natural gas, gasoline, heating oil, tar sands, crude oil, liquefied natural gas, diesel, antifreeze, fuel oil, polyethylene, styrene, coal bed methane, unleaded gasoline, propane, sulphur, hydrogen sulphide, and sour gas—the list goes on and on.

Oil is central to our lives, and it comes in many different forms. In its thickest form, oil is almost as hard as rock. In fact, most of it is trapped in rocks, buried far below the surface of the earth, stuck in geological formations. When oil gets warmer it becomes gummy, like molasses sitting in the sun. It is thick and sticks to your shoes. Gummy oil is useless for anything except fixing a hole in a bucket or greasing bearings. Warmer still, oil becomes a liquid that can burn. The gasoline we put in our cars is a clear, smelly, and very flammable liquid, so we treat it carefully, even as we use it every day in our cars, trucks, trains, airplanes, and ships. Heated up further, oil becomes an invisible vapour that is very flammable, like gasoline, and deserves the same respect. We burn it as natural gas in our home furnaces to keep warm in winter and fire it in generators to make electricity. Some cars and trucks run on it. Propane is a form of natural gas and is used in many of the same ways.

Oil can be a solid, a liquid, or a vapour.

Still confused? Don't worry. Though we use petroleum products in many different ways, they all come from the same source: hydrocarbons.

So far we've only talked about petroleum as fuel. It's also the source of many other products we use every day: medicines, plastics (like the siding on your house or the bumper on your car), the varnish on your coffee table, pencils, computers, running shoes, hair care products, and hockey sticks.

An Overview of Oil

This book helps answer your questions about oil and gas. Feel free to jump around, from chapter to chapter, to the index and the table of contents. The glossary will help you understand the difference between bitumen, tar sands, and oil sands—actually, it's all the same thing! You can also search the Internet or go to the library for further information on any topic covered in this book.

By reading this book, you can find out where oil comes from, how it got there, why it's trapped in the earth, and how we found it. Though oil is found many places in the world, as chance would have it, it's usually far away from where we need it. For centuries, wood and coal were the main fuel for industrialized economies. This book explains why oil and gas took over as the fuel of choice for many modern societies, and why we still use wood and coal.

Though it sometimes shows itself at the surface, petroleum is usually hard to find. Geologists, geophysicists, and many other experts search for it every day. They use increasingly complicated theories to understand how oil got trapped underground. Powerful computers help them model, or guess, where petroleum is hidden deep underground and under the sea.

Even after we find oil, we still have to get it out of the ground. We drill for it, mine it, force it to the surface with water, heat it with hot water and even fire—anything to get it to move from the rocks where it's trapped.

Once the oil is above ground, we have to process it. Most oil and gas is impure, mixed with dirt, rock chips, sand, sulphur, salt water, and dozens of other impurities. Refineries and processing plants upgrade raw petroleum into products we can use.

Hockey sticks and coffee tables contain oil.

Pipelines, trucks, trains, and huge ocean tankers then move oil and gas to our homes, gas stations, and the manufacturing plants that make thousands of products. The petroleum industry creates millions of jobs around the world. Drillers, ship captains, gas-station cashiers, computer designers, housebuilders, car salespeople, and thousands of others make their living off oil and gas.

Because it's an internationally used commodity, petroleum is also strategically important. Politicians and economists watch the price of oil every minute of every day. Wars, religions, weather, tides, consumer trends, and recreational habits all influence the price of oil. Oil company executives want a piece of the profits for their investors. Prime ministers and presidents want a piece of the action, too, as do religious leaders. Rebels seek control of oilfields in Colombia and Mexico, in the Middle East and Africa. The people who have access to oil have power.

Power is not easy to share. The leaders in the Middle East have one set of priorities—selling oil at a healthy profit. The leaders in the United States have another set—security of supply and a low price. Mexicans have their wishes, as do the Chinese and the people of Japan. Even the member countries of the European Common Market don't all agree on oil policy, but they try to work together in harmony.

Booms and busts are cyclical— about every ten years.

It's no shock that Canadians don't agree on petroleum policy either. When the price of oil was low in the past, politicians in producing provinces demanded access to markets and a set price *above* world levels. When the world oil price skyrocketed, consuming provinces wanted protection from rising costs and a set price *below* world markets. Then the price fell again and the rich provinces fell on hard times. Federal politicians tried to balance the economic needs of all parts of the country, without much success.

Booms and busts are also part of the petroleum story. They come, on average, every ten years. A war can cause a boom, or a bust. Shortages create booms. Recessions make busts. Ideologies contribute to the cycle too. For example, international business leaders want the free market to set prices, while Canadian politicians want more stability and try to control the economy through government programs. Native land claims can also affect the development process. During the mid-1970s the federal government delayed the Mackenzie Valley Pipeline until Aboriginal land claims and conservation issues could be settled. Technological advances, such as a new kind of car engine, can increase efficiency. Not enough pipelines, or too many, can affect the industry too. Environmental catastrophes, international agreements, economic downturns, terrorist attacks, hurricanes, or other regular but unpredictable events contribute to the booms and busts.

Oil is traded internationally, consumed locally.

Though most causes are international, the effects of booms and busts are felt locally. Housing prices skyrocket for a few years, then drop so far that people walk away from homes that are worth less than the mortgage, as people did in Calgary in the mid-1980s. Governments expand social services during good times and then feel forced to make brutal cuts during downturns like the mid-1990s. And then in 2005, the Alberta government gave a $400 cheque to every citizen as a bonus. The boom-bust cycle of the oil patch is hard to understand and even harder to predict.

It all seems so complicated, but it's really very simple. Petroleum comes out of the ground, and we use it in many ways. So let's go back to the beginning: oil is like water—it can be a solid, a liquid, or a vapour. ∎

Certain information with respect to this company contained in this presentation, including management's assessment of future plans and operations, contains forward-looking statements. These forward-looking statements are subject to numerous risks and uncertainties, some of which are beyond the company's control, including the impact of general economic conditions, industry conditions, volatility of commodity prices, currency exchange rates, competition from other explorers, stock market volatility and ability to access sufficient capital. The company's actual results, performance or achievement could differ materially from those expressed in, or implied by, these forward-looking statements and, accordingly, no assurance can be given that any events anticipated by the forward-looking statements will transpire or occur.–From an annual corporate presentation, 2007

Jargon

The Roughneck and
the Pump Jack

Once upon a time a broker, a rockhound, and a doodlebug talked a driller and his derrick into going into a play and wildcatting a prospect. First they used a cable-tool, but later they ran rotary and even went into diamond to get better cores and cuttings. After they hit a gusher and had a blowout, it turned into an inferno that had to be controlled by a firefighter. When they had run casing, they put on a Christmas tree and completed the operation; then they had to put on a pump jack when ROE lagged. Hydrates plugged the lines, condensate value fell, and deliverables went south, so they installed a compressor after deregulation and dispute resolution.

On the backhaul they saw another prospect and got the landman, named Jane, and the company hand in the doghouse to discuss the pool. OPEC was a problem, so they consulted the NEB and EUB and decided to go horizontal. Sour gas was a concern, so they used logs from orphan wells to access source rock. Vibroseis helped, too, and the proppant held open the tight formation so the wet gas and naphtha could get to the trunk line.

Some terms are just silly, others historic.

The next daylight tower the derrickhand and the leadtonghand went up to the monkeyboard to make up a joint, but later had to go open-hole fishing and eventually used a macaroni string. Crude and gasohol cannot both come to the wellhead at the same time, and in the end they had to kill it, but not before they had to use a BOP and bail. They also lost time when they steam-flooded the area and tripped out. When they felt trapped they used a scratcher and stimulation which resulted in a Quad, and the saddle plant took off the NGLs midstream. The greenhouse effect and VOCs were a concern, but the landed cost of the tight gas was in the public interest.

They did have one dry hole where they had trouble with junk even though they used mud. The operator wanted good downstream, but the packers did not co-operate. The jerk line, neutron log, and rathole were also a problem, and in the end they were spent in the process, so they moved on to a new unit.

What?

Every industry has its expressions, abbreviations, and funny words. The oil patch has got to take the cake—or the pool. Okay, so a **pool** is where oil or gas sits, trapped in a rock **formation**, far underground. But who was the roughneck, and how did the pump jack get involved?

A **roughneck** is a person who works on a **drilling rig**. The term "roughneck" or "redneck" comes from the fact that the

2

men, and they are usually men, get rough or red necks from working in the sun.

The **pump jack** is a piece of **oilfield** equipment. It bobs up and down like a kid's toy—one that looks like a bird dipping its bill down into water and then bobbing back up. But the pump jack is actually a pump. Its rod goes down hundreds of metres or more into the ground and sucks up oil from the bottom of the well. Most oil wells only produce a few barrels per day, and some of them need a pump jack to bring it to the surface.

Viking No. 1 gas well, Viking, Alberta, circa 1919.
(Glenbow Archives, NA-1072-16)

As you watch the news, read the paper, and drive through the countryside, you sometimes hear or see things that make you wonder about these and other oil patch terms. Some of the words apply specifically to something that happens on a rig or a pipeline, and make sense in context. Rigs have **doghouses**, which are shacks for the crew. Pipelines have mechanical **pigs** that go through the line and clean them out from the inside.

Others are just silly. The **monkeyboard** is a platform high up in a rig where the **derrickman**—the guy who works up near the top of the rig—stands and handles the top of the pipe. No monkey ever worked on a rig, of course. (Don't believe the story about the guy who got fired when the **driller** realized his pet monkey was a better derrickman. A week later he got a phone call. "Come back to your old job. Your monkey's drilling!")

In the Beginning

Petroleum is found in **pools** deep underground in **traps**—places where rocks keep it stored. Oil or gas can be in a tight formation or rock that is not willing to give up the petroleum in the tiny holes in its structure. In some places **salt domes** hold the **reserves**, in other places they are in coral reefs, old riverbeds, and many other kinds of formations. **Geologists** have assigned them many names, such as the Devonian, Mississippian, Sekwi, Colorado, Peace River, and Turner Valley. Sometimes the names refer to the age of the rock, and other times they relate to the nearest town, river, or other geographic feature.

Crude oil is unrefined, not a poor host at a party. **Refined oil** is processed in a **refinery** and what we use to lubricate the engines in our cars. **Natural gas** can contain **ethane, methane, propane, butane, pentane**, and many other "anes," and they are all gases. Natural gas can also contain contaminants like water, salt, **sulphur**, and many other products that come with it from the rocks deep underground.

Naphtha means "liquid bitumen" in Latin.

Some of these contaminants are harmless, but still have to be removed from the gas. Others are deadly, even in small quantities. **Sour gas**, for example, contains **hydrogen sulphide** that, once removed from the gas and processed, becomes elemental sulphur, a product used by many industries. When cleaned of its **impurities**, sour gas becomes **sweet gas**. **NGLs** are natural gas liquids, the liquids that are usually found with gas. Old-timers called this product **naphtha**, and there is a little hamlet in southern Alberta named after this liquid. In the early days most oil companies just burned off **waste gas** at a **flare**. It was called "waste" gas because there were no markets for the vapours that came out of the ground with the oil or natural gas liquids.

*Drilling is upstream, refining is midstream,
and the gas pump is downstream.*

Upstream: The Scientists

People who work in what is called the **upstream** end of the industry find the oil and gas. This is the first part of the process that moves oil from deep underground to your car.

Even though it has many names and uses, petroleum is not much use to us when it's buried far underground, so **explorationists** go looking for it. Geologists use the science of geology—the study (*ology*) of the earth (*geo*)—to understand the layers of rock, soil, clay, and other materials that make up the top few kilometres of the earth's crust. Geologists do **surveys** and **map** features, all the time trying to see below the surface. Imagine a birthday cake made of many layers of different cake—chocolate, vanilla, carrot cake, marshmallows—with icing between them and candy or quarters hidden between the layers, and you get an idea of what geologists do. They look for the money—the oil.

Another kind of scientist who helps in this quest is the **geo-physicist**, literally, a person who studies the physical ways the earth works. Today's geophysicists are highly respected and well-educated members of every team of explorationists who look for oil and gas and many other minerals. They apply complex mathematical ideas and instrumentation and computers to the quest for oil. They are also working on the same birthday cake, but they bounce **seismic**, or sound waves, through the cake and listen for hints as to the location of the marshmallows and coins. A key person on an early geophysical crew was a **computor**, the skilled mathematician who did the computations in the days before computers. He made sense of all the **records** collected from the seismic sound waves and stored them on photographic paper as squiggly lines. Perhaps they were called **doodlebugs** because of the doodles on the paper. Actually, this funny term probably came from the dowsing they did to find oil—**dowsers** have been finding water with forked sticks or other kinds of rods for hundreds of years.

Upstream: At the Wells

After explorationists identify a likely spot to drill, the driller actually finds the oil by sinking an oil well. A **drilling rig**, or **derrick**, can be made of wood, but metal rigs are most common today. Like the electric drill you use to make a hole through a board, or your electric mixer, a drilling rig turns a bit and pushes its way through obstacles. Soil, sand, clay, soft rock, water, and very hard rock can stand in the way between the drill bit and the pool or **reservoir** that will make the well a **producer**. Sometimes it's a **dry hole**—a failure. **Wildcat wells**, drilled in untested areas, are often dry, but even a "failure" can contribute valuable knowledge to the geological understanding of an area.

Carter Oil seismic explosion near Lethbridge, Alberta, 1936. (Glenbow Archives, NA-3833-2)

Casing in the hole keeps the sides from falling in. All kinds of **bits** or **tools** can be used in the drilling operations, depending on the obstacles found along the way. A company can order **directional drilling** (drilling on an angle) in order to drill several wells from one location, or to get to a spot under a mountain, lake, or river. **Horizontal drilling** (drilling parallel to the surface) is another option and often is used when the well hits **production**, or petroleum, and the reservoir is thin but extends laterally from the well or does not have enough pressure to push the oil up to the well.

Sometimes you will see a **service rig** at an oil well for a few hours or even a day or two. It is doing repairs to the well or some other procedure such as **acidizing** or **perforating** the well in order to increase production. A service rig maintains a well and keeps it operating efficiently. **Secondary recovery** involves applying water, gas pressure, or other techniques to a well or an oilfield in order to make it more productive. **Mature fields** are not just old, they are also in decline, and so companies expand the quest for oil to new **frontiers**. **Offshore drilling** involves drilling for oil and gas out at sea using **drill ships** or **drilling platforms** in regions such as the continental shelf off the east, west, and northern coasts of Canada.

Imagine an oilfield as a big balloon, under pressure, under a couch cushion at a birthday party. Kids jump on the couch and move the cushions around and maybe even tip it over. With enough roughhousing they can even break the balloon, and all the air will seep out, or escape with a pop. An oil well **blowout** is like a pop in that balloon, sometimes creating a **runaway** or a **gusher** of oil. If it catches fire an **oilwell firefighting company** will have to be brought in to put out the blaze.

For the most part, oilfields are not one big cavern thousands of metres down, waiting to expel their oil and gas or to be

Atlantic No. 3 well blowout near Leduc, Alberta, 1948.
(Provincial Archives of Alberta, P2824)

sucked dry. Most **plays,** or operations, are complex. Rock can be **fractured** or **faulted** or contorted, with layers thrust over or under other layers, and the whole area can be tilted or cut through by underground or surface rivers. Oil, gas, liquids, salt water, **sweet water,** and many other impurities can come up out of the ground together. These all have to be handled and disposed of in an environmentally appropriate manner. **Productivity** is a challenge, and even the most successful oilfields keep most of the oil and gas firmly trapped in their rocks. The Turner Valley oilfield, discovered in 1914, has released less than 10 per cent of its oil and gas. Even though many different techniques have been applied to get it to give up more, total recovery from this field will probably be less than 15 per cent.

Midstream: Pipelines and Processing

The **midstream** part of the industry upgrades and moves oil and gas products to markets, such as gasoline stations, homes, and factories. Once the product is out of the ground, it has to be processed. **Field lines** collect oil from numerous wells and move it to a central location. It then moves by large continental pipeline to where consumers live: Ontario and Quebec in Canada, or the American Midwest or California. Some of these lines go by the names of IPL (Interprovincial Pipe Line), West Coast Transmission, and TransMountain Pipeline. Most of them date back to the 1950s and 1960s when supplies of oil and gas made long-distance **transmission** economically viable. One of the earliest of these megaprojects was called the **Canadian American Norman Oil Line (CANOL)**, a World War II pipeline that took oil from Norman Wells on the Mackenzie River in the Canadian Northwest Territories to a refinery in Whitehorse in the Yukon, 950 kilometres (590 miles) away. It moved oil for less than a year, cost more than $130 million to build, and delivered under a million barrels of oil in 1944 and early 1945—at about $138 per **barrel**. It was far too expensive to remain in business after the end of the war, but its story illustrates the expense and risk associated with building pipelines to move petroleum to markets during a temporary economic or political crisis. The **Mackenzie Valley Pipeline** is the most recent of these expensive projects. It was postponed in the 1970s after the **Berger Mackenzie Valley**

Pipeline Inquiry, but will eventually bring massive quantities of natural gas out of the Arctic, one of Canada's most promising frontiers. Oil and **liquefied natural gas (LNG)** also come to Canada by oil tanker or LNG tanker from other oil-producing areas in South America, the Middle East, and Europe.

Welding a pipeline, circa 1930s. (Provincial Archives of Alberta, P1973)

Natural gas is cleaned, then made smelly again.

After arriving by tanker or passing through a transmission pipeline, oil goes to a refinery where it is **fractionated**, or boiled in a tower, and the heat and pressure force off various compounds: grease for lubrication; **heating oil** for use in furnaces; **kerosene** for burning in camping lamps, and a similar product called **aviation fuel** that jet engines burn; **diesel** to fuel trucks and railway engines; various grades of gasoline; household products like **turpentine** or paint thinner; and many, many others. Natural gas goes to a gas-processing facility, or **gas plant**, where impurities and products like butane and sulphur come off the gas. All smell disappears in the process of making the gas sweet, so workers add **mercaptors** to the gas so that it gives a warning smell to utility workers and warns consumers if they are exposed to a leak. This part of the process is called the midstream, the part between the oil in the ground and the consumer.

What You Can Find in a Barrel of Oil

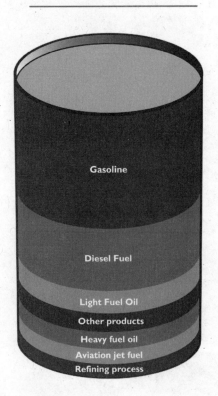

Gasoline

Diesel Fuel

Light Fuel Oil

Other products

Heavy fuel oil

Aviation jet fuel

Refining process

Downstream:
Delivering to the Consumer

The **downstream** end of the industry is the part that sells products directly to the consumer: gasoline, plastics, fertilizers, and toothbrushes. Almost half the petroleum produced each year goes into manufacturing plastics, medicines, cars, construction materials, and other consumer products, so that part of the process is invisible. We buy gasoline at the gas pump at a service station, which was named after the service that they once provided to the automobile. This service included things like changing the oil in the engine, replacing the tires, doing tune-ups to the motor to keep it running efficiently, and working on the transmission and other parts of the power train that make the car stop and go. But today most of us go to self-serve gas stations to buy convenient (and expensive) snacks, to clean the windshield, to use the bathroom, and to buy 50 or 75 litres (13 or 20 gallons) of gasoline for the tank in the car. We seldom notice that the bottled water in the snack area costs more per litre than the gasoline.

The other main use of petroleum is as natural gas to heat our homes. Furnaces and water heaters burn natural gas as a fuel, and it comes by **low-pressure pipeline** to our homes from a **residential gas company**. In some parts of the country **fuel oil** is more common in home heating, but natural gas is a cleaner-burning and usually cheaper fuel, thereby making it the product of choice. We measure oil in a barrel by liquid quantity—192 litres (51 gallons), and one cubic metre of oil, another international way of measuring crude, is about six barrels, or 1,200 litres (317 gallons) of oil. We buy natural gas by gas volume at a specific pressure, so many **MCF** or thousand cubic feet.

Pricing of petroleum is done according to an international standard because oil and gas are sold as global commodities. The **West Texas price** for oil is often mentioned because it is a **benchmark crude**, or a standard quality of oil against which other products are measured. In other words, a barrel of oil of a specific quality today is worth $67.23 in West Texas. The daily quote for the **price of crude oil** is a relative number, with value being added to and removed from each barrel of oil depending on its quality, its distance from markets, **international commodity speculation**, and numerous other factors, including weather projections for next winter, the hurricane season, inflation, recessions, the spending and borrowing habits of people around the world, political instability, war, and the sales of suvs (sport utility vehicles).

Ownership and Control

Thousands more terms are available online at industry association websites like the one maintained by the **Canadian Association of Petroleum Producers (CAPP)**. As with many other terms in the oil patch, CAPP is not exactly what it seems. It is, in fact, an association of the largest and most powerful oil companies *operating* in Canada, and its members produce over 90 per cent of all the oil and gas in the country. But, for the most part, they are not Canadian companies. Most are owned by large multinational corporations with headquarters in the United States, England, Holland, and elsewhere, but not in Canada. Is this important? Some people consider the source of capital and control over the Canadian petroleum industry a sovereignty issue. Others don't care where the money comes from or who controls the industry, as long as Canadians are prosperous and have lots of work. As this book shows, the oil patch has always been a fiercely political industry, both locally and internationally. It's up to you as a voter to decide whether or not Canadians should own and control this important part of our economy. ■

In case you're still wondering, the words in **bold** in the first paragraphs of this chapter will all be explained in the text of this book and the glossary. **Vibroseis,** for example, is a way of sending a vibration down into the earth, instead of using dynamite, so that seismic recorders can pick up signals bouncing off rock formations far underground.

Silly words, words used out of context, and abbreviations are common in the Canadian oil patch. And they're not always what they seem! ∎

Key Dates

1719 –   Oil found leaking into the Athabasca River

1820 –   Kids in Nova Scotia made flares at holes that leaked natural gas

1830 –   Oil noticed in swamps at Petrolia, Ontario

1874 –   Crude oil noticed in the Waterton area of Alberta

1883 –   Natural gas found at Langevin near Medicine Hat, Alberta

1914 –   Turner Valley oilfield found southwest of Calgary

1919 –   First well hit oil on the Mackenzie River in the Northwest Territories

1947 –   Crude oil discovered at Leduc, Alberta

1966 –   First oil well test in the Beaufort Sea in the Far North

1979 –   Hibernia drilling discovers oil off Newfoundland

1985 –   Norman Wells Pipeline begins moving oil out of the Canadian North

Discovery

Oil Is Where
You Find It

The horses reared back from the bubbling water, almost tipping the wagon over sideways into the river. "Whoa," yelled the teamster. With a crack of the whip he moved the horses ahead, up to another part of the creek bank. Safely past whatever it was that had spooked the horses, the rancher stopped for lunch.

He'd been coming up the Sheep Creek in southern Alberta for years to get coal. It wasn't much of a mine, really—just a hole in the side of the riverbank. The Okotoks rancher used coal in his house and in the blacksmith forge when he worked on horseshoes.

But this day was different. The horses were skittish, frightened of something. Was it that bubbling water? Bill had never seen anything quite like it, although it always did smell funny along this stretch of the river.

Bill sat and ate his lunch. When he lit his pipe afterwards, he threw the match into the bubbling water. It exploded! Not a big boom, but a sharp crack. The water roiled and surged.

What was this all about? Sometimes the creek had a bit of a sheen on it too, and he'd heard about swamp gas. Next time he'd have to bring a jar along and see if he could catch some of this smelly stuff.

That's how Bill Herron of Okotoks, just south of Calgary, ran into petroleum in about 1911—natural gas, bubbling up through a pool of water alongside the creek. To this day, when it rains and water collects on top of the soil at that spot, you can still see gas percolating through the water. In 1914 Bill's oil company, Calgary Petroleum Products, developed a successful oil well on that site.

Oil Pops Up in Strange Places

For generations Canadians have been finding oil from one end of the country to the other, usually far from urban areas where large numbers of people might actually use oil and gas products.

In 1719 a Cree man named Wa-Pa-Su took a sticky lump of oily sand from the Athabasca River to Fort York, where he presented the mess to Henry Kelsey. The Hudson's Bay Company official was unimpressed, but, being a fastidious record keeper, he mentioned in his diary on June 12, 1719, that he had received a sample of "the Gum or Pitch that flows out of the Banks of that River."

In 1778 Peter Pond noted "springs of **bitumen** that flow along the ground." A feisty American explorer of ill repute, Pond encountered the pitch while travelling down the

Clearwater River to where it flows into the Athabasca, near today's Fort McMurray.

In the Maritimes people noticed natural gas in the early days too. Children near Lake Ainslie on Cape Breton Island in 1820 had fun playing with fire. They drove stakes into the soil to make holes and then lit the natural gas that escaped from them, making little flares. In the 1840s Dr. Abraham Gesner investigated **oil shales** at Frederick Brook, not far from Moncton, New Brunswick, and from the 1850s to the 1870s miners dug it out of the ground and sent it to Philadelphia and Boston, where refiners made it into kerosene.

In 1859 Dr. H. C. Tweedle found both oil and gas on the Petitcodiac River near Dover, New Brunswick, but water seepage prevented production from four shallow exploratory wells in the Dover field. Drillers also found oil across the river in 1909 and named it the Stoney Creek field. By 1912 a pipeline was delivering natural gas from this field to Moncton.

Chance discoveries also happened in central Canada. In southwestern Ontario, local Natives had long known about gum beds, where oil had seeped up out of the ground and pooled on the surface. They used it as medicine and to pitch their canoes to make them waterproof. According to Victor Ross, an early Canadian oil patch writer, Europeans knew about oil. Ross writes in *Petroleum in Canada*, "In 1830 the settlers in the vicinity of Enniskillin, in Lambton County—the extreme western part of Old Ontario—noticed the presence of oil in the swamps of that region. It was called 'gum oil' and was present in such quantities as to be regarded as a nuisance because it killed vegetation and so detracted from the value of the land."

As people moved into other parts of Canada they ran into petroleum there too. They made small discoveries of natural gas at Pelican Point on the Athabasca River in northern Alberta in 1898, in southern Alberta at Langevin in 1883, at Medicine Hat in 1890, at Bow Island in 1909, and at Turner Valley in 1914. The newcomers began developing crude oil in today's Waterton Lakes National Park in Alberta in 1901 and on the shores of the Mackenzie River in 1911.

Drill Where?

Waiting to find oil by accident was not good enough, so by the early 1900s oil companies were sending geologists out to find it scientifically. In 1919 geologist Ted Link, a drilling crew, supplies, and an ox named Old Nig departed Edmonton, heading north. Six weeks later they arrived at the mouth of Bosworth Creek near oil seepages on the shores of the Mackenzie River. They had just

Bill Hill, Ted Link, Elmer Fullerton, and W. Waddell at Bear Island, Mackenzie River, Northwest Territories, 1921. (Glenbow Archives, NA-463-24)

Oil Creek, Northwest Territories, 1921. (Provincial Archives of Alberta A-11312)

enough time to set up camp and build a wooden drilling rig before winter set in. Legend has it that Link chose the site by waving his arm and saying, "Drill anywhere around here."

Drillers ate the ox when they ran out of meat.

Though at times flippant, Link was a superior scientist and eventually became Imperial Oil's chief geologist. The ox, however, was not as fortunate. Old Nig went into the cook pot during the winter when the men ran out of meat. Another drilling crew arrived in the spring of 1920, and in August they hit oil at a depth of about 1,240 metres (4,068 feet), the world's most northerly oil well to date. The well produced 600 to 900 barrels per day at first, but settled down to an average of about 100 barrels daily. Unfortunately there was no one there to buy the oil.

At this time geologists started to look for oil closer to markets. In 1861 Thomas Sterry Hunt theorized that petroleum deposits were related to plants. Oil is usually trapped in porous rocks, under layers of impermeable rock that work as traps. Where old seas deposited sediments on top of reefs, for example, decaying plant and animal matter was trapped. Heat, pressure, and decay eventually turned the organic material into oil and gas, or coal.

Harry G. Lloyd, Garrett Greene, A. P. "Tiny" Phillips, and W. R. "Frosty" Martin at first gas well called "Old Glory" at Bow Island, Alberta, 1909. (Glenbow Archives, NA-1072-1)

What They Do...
Geologists and Geophysicists: Squeezing Oil Out of a Stone

Petroleum geologists and geophysicists play with rocks all day. Their job is to imagine what is going on deep in the earth today, as well as what happened thousands of years ago. In the early days they had to guess, using rock outcrops and riverbanks to give them an idea as to what lay far below the surface.

In the past, geologists spent their summers in the field, collecting rock samples and making maps. These days they spend most of their time working on computers. Geologists also study math, chemistry, or biology to help them look for oil and gas.

Geophysicists learn about the composition of the surface and the depths of the earth. They study watercourses, magnetic and gravitational fields, as well as other physical principles, and use powerful computers and seismic data to map and model the places that might hold black gold.

When theory, research, and hard work combine with some luck, they find new oil and gas fields. ∎

That was the theory. To test it, geologists had to find places where petroleum was trapped. Gas bubbles away quickly, and even oil tends to move, floating away on top of water.

Canoeing was a popular way of looking for oil.

Early geologists started going on canoe trips to look for traps. Field crews with the **Geological Survey of Canada (GSC)** and most major oil companies went out every summer, canoeing along the rivers and on the lakes in the North, looking for hints in exposed rock outcrops. Soil, trees, and shrubs cover much of Canada, but rocks break out along mountain ranges and beside

Jack Porter taking core at Canadian Superior Jones well, Virden, Manitoba, 1950.
(Glenbow Archives, NA-3833-1)

rivers. Geologists also spent summers climbing mountains and making careful records of all the layers of rock and the angle of tilt, taking samples. In the winter they mapped these features and made educated guesses about what was happening far under the surface. Huge regional survey and mapping projects helped them understand the formation of the earth.

Geologists also learned from drillers. Most early successful oil and gas wells were located on the basis of a surface seepage, but each time the drillers emptied the cuttings—pieces of rock, clay, or dirt—from the well, they collected samples for the geologists.

Careful records helped them understand the layers of different kinds of rock through which the drill bit cut as they bored hundreds or thousands of metres deep into the ground.

Cuttings from a few wells told the geologists very little, but samples from hundreds of wells taken in the Turner Valley oilfield in Alberta allowed scientists to create an elaborate model of the subsurface geology. Endless speculation by the geologists

The February 13, 1947 gush of the Leduc No. 1 well marked the beginning of Alberta's post-war boom. (Provincial Archives of Alberta, P2722)

gave them lots to do, but the real discoveries in the 1930s came from stepping out—that is, drilling new wells beside other successful producers—so the geologists learned from the drillers who maximized the potential of the Turner Valley field, and by 1930 Alberta was producing more than 90 per cent of all the oil in Canada.

Though Turner Valley was an important producer, it was a one-of-a-kind oilfield. Companies ran up and down the east side of the Rocky Mountains looking for other oilfields like it, but there were no more. Instead, the next major oilfield was hiding far out in the plains, hidden under farmland.

Ted Link's oil well up on the Mackenzie River had found oil that was trapped in ancient coral reefs, but geologists considered it an **anomaly**, also one-of-a-kind. But the drill bit—the only reliable "oil finder"—made petroleum history on Friday, February 13, 1947. Imperial-Leduc No. 1, drilled just south of Edmonton, discovered a new oilfield and changed the geological understanding of petroleum.

By the end of 1947, 28 of the 33 wells drilled in Alberta's newest oilfield were flowing oil! The Leduc-Woodbend oilfield's production surpassed Turner Valley in 1948, and by the end of 1951 oil production from it and Redwater was twelve times greater than the rapidly depleting old-timer at Turner Valley. Its production had peaked at just over 27,500 barrels per day in 1942, while Leduc and Redwater combined pumped out almost 40 million barrels in 1951, or more than 100,000 barrels per day.

The new discoveries in the prairies also found enormous quantities of natural gas—far more than Alberta could use. Most of the gas in Turner Valley had been flared while producing the oil, so in the mid-1950s TransCanada PipeLines Limited built the first of many large gas pipelines to distant markets. And with new markets to supply, oil companies had to find more oil and gas.

The hunt was on.

Forked Sticks, Witching Sticks, and Black Boxes

Rock samples, maps, and drill bits were not the only way to look for oil. Men with forked sticks or bent wires had been dowsing water wells for many years. Maybe their success was based on the magnetism of the earth, or perhaps on the dowser's

Imperial Oil seismic recorder, Peace River, Alberta, circa 1940s. (Imperial Oil, 9807)

intuition or his subconscious knowledge of the terrain and the most likely places to find underground rivers.

Some people used forked sticks to find water and oil.

Dowsers started offering their services to oil companies in the 1920s. Others showed up in Alberta with oil "detector" boxes. One, the Mansfields Water and Oil Finder, promised success. At a location east of Calgary at Gleichen it "found" gas at 216 metres (708 feet), gas at 457 metres (1,500 feet), gas and oil at 670 metres (2,200 feet), oil again at 826 metres (2,710 feet), and oil at 1,039 metres (3,410 feet). But an oil well drilled on the basis of this information found nothing.

By the end of the 1920s some of these methods were becoming accepted by oil companies. The *Imperial Oil Review* in 1928, for example, reviewed current methods for finding oil, including electrical, "gravimetric, seismic and magnetometric" techniques. Though these all sound like far-fetched ways of duping the public, the article was based on research conducted by Dr. Conrad Schlumberger, professor of physics at the School of Mines in Paris.

Conrad and his brother, Marcel Schlumberger, were pioneers in geophysics and invented the "**electrical logging**" technique that became common in the 1930s and is still used today. Along with other tests done in oil wells, electrical logging tests the electrical conductivity at different levels in an oil well. It can help identify different kinds of rock, clay, sand, and other materials. Like rock samples taken by drillers, it can also help map the well and the surrounding area.

Seismic also became common in the 1920s and 1930s. A small, truck-mounted drilling rig makes a series of holes in a line while dynamite charges in the bottom of them make an artificial earthquake. Sensitive listening devices called **geophones**, like headphones, capture the sound of the explosion and record it. In the early days the records were squiggly lines on photographic paper, but today's seismic records are huge computer files.

Canadian-built tracked vehicles can almost walk on water.

As the vibrations from a seismic explosion move away from the **shot hole**, they hit layers of rock, clay, or sand and bounce back. A trained geophysicist can read the recordings and, like a geologist, create a map of the layers far below the surface. Though seismic reflections don't identify oil or gas—they only bounce off hard objects—the maps based on the information they provide can indicate a trap. Only an exploratory well can actually find oil or gas, but geologists and geophysicists work together closely to understand the formations that store and trap petroleum.

Since the discovery of petroleum far out in the prairies in the 1940s, explorationists have expanded their search to all parts of Canada and all over the world. It is hard to travel in many parts of northern Canada in the summer because of boggy conditions, so seismic crews wait until winter freezes the land. Early tracked vehicles left over from World War II were modified to carry drilling and recording equipment. Canadians eventually designed and built some of the best tracked vehicles, the Nodwell and Foremost equipment, for example, and used them in our North and exported them around the globe.

Much of Canada is underwater, so exploration programs have taken to rivers, lakes, and the oceans in the search for petroleum, too. Since the 1950s, seismic explosions in the

29

What They Do...
Landmen: Lots of Cups of Coffee

Landmen talk to people all day and get them to sign papers. Male and female landmen are negotiators. They make the deals that keep landowners and companies getting along.

Surface rights and **mineral rights** are both required before a company can gain access to the land and permission to search for oil and other minerals far below the surface. Government permits are also necessary. In addition, landmen get permission for a company to install a pipeline or construct a refinery or gas plant.

Today's oil patch is a very busy place, so landmen are always busy communicating with many parties and applying research and negotiation skills to make deals and keep the complex oil industry getting along with members of the public. ∎

Mackenzie River in the North have helped create signals to map the geological formations in that frontier. Ships also conduct research programs in lakes and off all three coasts—the Pacific, Atlantic, and Arctic—making significant oil and gas discoveries in all these frontier regions.

Explorationists create 3D underground maps.

Today, complex geology and geophysics use the records from previous mapping, sampling, and testing programs, and drilling and production records, as well as new exploration programs, to continue the quest for oil. Seismic techniques have also become much more sophisticated, often using vibrations instead of explosions to minimize the impact on the environment. Helicopters sometimes transport the equipment, speeding up the process and limiting the damage caused by wide swaths cut through the forest for heavy-duty vehicles. Computers can collect vast amounts of information and process it in **3D**. A scientist at a computer workstation can use a joystick to move around, deep underground, exploring in a world of virtual reality on the computer screen in the quest for oil.

Today's discoveries are different from those of the past. Almost all the big oil and gas fields were discovered decades ago, when they were easier to find. Recent discoveries are smaller, hidden underneath previous oilfields, or in remote areas that were not economically viable until the price of oil rose. Today's frontiers are also in the understanding of economics, politics, and cultural and social values. The long and complex negotiations that were required to build the pipeline down the Mackenzie Valley out of the North are typical of this new era of oilfield development.

The quest for oil in Canada has gone from the simple to the complex, from accidental to intentional. At first we just drilled where oil and gas oozed or bubbled up out of the earth, but then we learned to look for it, to try to understand the geological record. Human history is short compared to the story of the earth, but scientists have learned to see its clues and interpret the traces left through time. As the search for more oil and gas continues, scientists are constantly looking for more understanding and applying new techniques and technology. ■

Key Dates

Total Wells of All Types in Oil and Gas in Canada

Year	Number
1947 –	431
1960 –	2,536
1970 –	2,936
1975 –	4,032
1985 –	11,750
1990 –	5,765
1995 –	11,062
2000 –	18,480
2001 –	19,752
2002 –	17,182
2003 –	23,365
2004 –	24,874
2005 –	28,353

(Source: Daily Oil Bulletin)

Ways of Releasing Oil from the Ground

Shovel –	10 metres per day
Jerker-type drill –	30 metres per day
Cable-tool rig –	100 metres per day
Rotary rig, circa 1950 –	300 metres per day
Modern rotary rig –	1,000 metres per day

Gushers
Drilling for Oil

D ig a hole and find oil—that's all there is to it, really. There are many ways to dig a hole, and most of them can find oil.

Though the Americans like to claim they were the first to find oil in North America in 1859, we Canadians are the record holders. Back in 1858, J. H. Williams of Hamilton, Ontario, took a shovel and dug down about 15 metres (50 feet) to create the first oil well in North America. This is how the famous "gum beds" of Oil Springs, Ontario, began producing oil. The distillation process "produced a comparatively light, iridescent liquid." Williams dug deeper and found that production increased with depth.

There Are Many Ways to Dig a Hole: From Spring Poles to Cable-tools

Digging holes in the ground with a spade is slow work, so in the early 1860s, when Hugh Nixon Shaw starting drilling for oil at

Oil Springs, he used something different. Shaw's drilling apparatus was primitive, just an 8-centimetre (3-inch) spring pole, but he was lucky. Finally, on January 16, 1862, he hit oil at a depth of 48 metres (158 feet) into the rock.

The oil rushed up Shaw's 7.5-cm (3-inch) hole, filled the 1.2-by-1.5-metre (4-by-5-foot) well dug through 15 metres (50 feet) of clay, and overflowed at the surface, a great, black, bubbling, and gurgling spring of oil. The oil spread everywhere, shot up in a gusher for about 6 metres (20 feet), and flowed at a rate in excess of 2,000 barrels per day, covering the ground for acres.

The **spring pole drilling system** Shaw used was very simple. You take a springy pole about 10 metres (33 feet) long, bend it over a fulcrum of some sort—a crotch in a tree works fine, as does a big post with a V cut in the top—and then fasten the other end to the ground. A rope tied to the other end is the drilling line, and a heavy weight with a sharp end at the bottom serves as the drill. Then you pull down on the end of the spring pole to get the drill bit to cut into the bottom of the hole and let the natural spring of the pole pull it back up.

Up and down, up and down, hour after hour. A day of hard work drilling with this system might produce less than a metre of hole! Spring pole drilling systems evolved over time and eventually used horses, walking on a treadmill, to provide the power. But drilling with a spring pole was slow, and the wells never went very deep—maybe 75 metres (246 feet) at the most.

Nonetheless, the technique showed what could be done with simple technology. The next development was the **cable-tool rig**, a system still used today for drilling water wells. Like the spring pole, it simply lifts and drops a heavy bit into a hole, and the force of the falling metal cuts its way through rock, clay, soil, and anything else it encounters.

Tall wooden derricks, 30 metres (98 feet) or more in height,

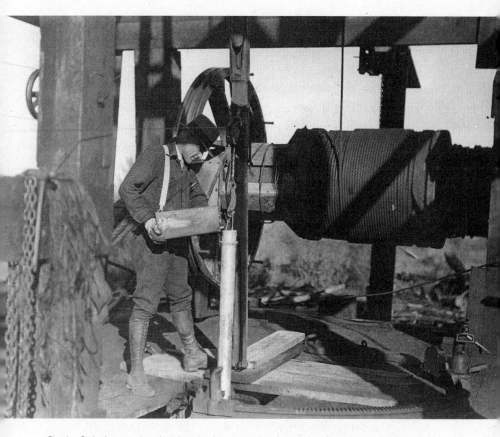

Charles Stalnaker pouring nitroglycerine into torpedo tube used to fracture an oil well and increase production of oil, circa 1935. (Glenbow Archives, NA-3682-45)

were part of the first cable-tool system. Built of large wooden planks and nailed together with spikes, they could be built on site by a carpenter or even the driller. A boiler provided the steam (wood, coal, or even natural gas was the fuel) for the engine that powered the belt that turned a band wheel. A pitman connected the band wheel to the walking beam, and this combination converted the circular motion of the band wheel to an up-and-down motion that lifted and dropped the drilling cable and the heavy metal drilling bit at the bottom of the well—just like the spring pole system but much more ambitious. Pulleys and other wheels pulled the drilling cable, made of hemp or wire rope, out of the hole when the cuttings at the bottom prevented further drilling. Then the driller dumped a barrel of water down the hole and lowered a **bailer**, a piece of pipe with a one-way valve at the bottom that opened when it hit bottom and swallowed up the **mud** and cuttings. Pulled out of the hole, the bailer dumped its waste into a chute that got rid of the cuttings. Typically, they made 3 or 4 metres (9 or 13 feet) of hole in a good day.

The first oil rigs were wooden, spiked together.

When all goes well, a cable-tool rig can drill a lot faster than a spring pole rig, but it often took many months or years to drill a well. Holes went crooked when they hit obstacles or sloping rock formations. The bit got stuck in the well, and the driller and his helper had to "**fish**" for the tools downhole. When all else failed, they sometimes had to abandon the hole, move the rig over a few metres, and start all over again.

Even when the bit went straight, the drillers had to install casing to keep the sides of the hole from caving in on the drilling operation. The casing was a length of steel pipe, 7 metres (23 feet) long, threaded so that it could be screwed together with the others in the hole. Drilling, bailing, and running-in casing were time-consuming and involved repetitive tasks. Drilling crews also had to collect samples from the cuttings, or bailings, that came up from the bottom of the well so that geologists could learn about the formations deep underground.

Even with cuttings, in the early days the geologists were guessing most of the time. Geological theory was as primitive as the drilling technology. Even when the well hit the target depth—cable-tool rigs seldom drilled deeper than 1,000 metres (3,280 feet)—without finding oil, there was still a chance that other techniques might make the dry hole a success. Cement then set the casing into place to keep it from moving in the hole or coming up out of the surface in case of a blowout. Sometimes a specialist would perforate the side of the pipe at strategic levels and put acid, **nitroglycerine**, or other chemicals into the formation to encourage the movement of oil to the well.

Rock is hard, but it can have small holes in it that contain oil or gas. Opening up the formation can allow the pressure of hundreds or thousands of metres of rock and soil above it to push petroleum up the well to the surface. Even a dry hole can be made productive with careful production methods.

But cable-tool rigs were slow. In 1921, for example, companies drilled about 50 wells in Alberta, but in 2005 they drilled approximately 28,000, and not with wooden rigs or thumping bits at the ends of long cables.

Even though cable-tool drilling was a slow process and hard to control, the oil industry in Canada relied on this drilling technology for decades as it developed oilfields in Ontario, Alberta, and the Northwest Territories. Well into the 1940s Canadian drillers became known around the world for their skill and ingenuity with these cumbersome wooden monsters.

Getting There Faster: Rotary Drilling

Today's rigs use rotary technology, like the electric drill you use to drill a hole in a piece of wood, but these rotary drilling rigs are huge and powerful and can drill as deep as 5,000 metres (16,400 feet) into the ground. This technology arrived in Canada in the 1920s from the United States.

In 1925, after the Royalite No. 4 cable-tool rig drilled through a limestone formation and discovered a new pool of sour gas, Imperial Oil imported two rotary rigs for use in the new "wet" gas field. The 35-metre- (114-foot-) tall rigs covered three train cars and were valued at $250,000. Though these early rigs were impressive, they had trouble drilling down through the tilted and

Drillers Martin Hovis and Joseph Brown on the floor of the cable-tool rig at Dingman Discover Well, 1914. (Provincial Archives of Alberta, P1303)

twisted rock formations in the Turner Valley oilfield, and for a time the rotary drilling technology drilled from the surface down to the limestone, and the slower but more reliable cable-tool units bashed their way through the complicated limestone formations. Two years later, in the *Bulletin* of the American Association of Petroleum Geologists, a summary of Turner Valley drilling technology confirmed that this drilling combination was the only reliable method of overcoming the severe technical difficulties posed by the Turner Valley oilfield. In 1937 the first natural gas-driven rigs arrived from the United States, and in 1939 Ralph Will of Drilling Contractors imported the first diesel-driven rigs for use in the Turner Valley oilfield.

Drilling oil wells in Canada is often more expensive than in the United States.

Drilling in Turner Valley was expensive. Imperial Oil figures in 1938 revealed an average cost of $212,000 for each well drilled, an amount almost three times the American average, but the rigs drilled faster than their cable-tool cousins, and soon the metal towers and rotary rigs were industry standard. The steel rigs could be moved too, unlike the old wooden rigs that were just left to rot or topple over during a windstorm.

In place of slow and awkward belts and wooden walking beams, the rotary rigs boast a revolving steel turntable that grips the drill pipe that is connected to the drill bit. The most common bit is a cone bit with several cones geared to work together, cutting into the rock at the bottom of the hole. Together they grind and scrape and chew their way deeper and deeper into the earth. There is no bailer in this system, though, and drilling fluid, often called "mud," is forced down the middle of the pipe and through small holes in the bit, flushing the cuttings up around the drill pipe and back to the surface. Mud also keeps the drill bit cool and helps coat the inside of the well bore so it does not need steel casing like the cable-tool hole. It can be almost as thin as water or stiffened up with clay and chemicals to make it as thick as gooey plaster, depending on what it needs to do to keep the well in shape.

The rotary well is also a good source of information for the

Crew member on rotary rig, Turner Valley, circa late 1930s.
(Glenbow Archives, NA-67-113)

company geologist, so a **drilling log** and samples of the cutting are kept and carefully analyzed in a lab. Well-logging was another tool, and early electrical logging equipment before World War II helped geologists understand the character of the rock through which the rig was drilling. Other instruments help geologists understand the angle of the drilling string. Although bits last much longer today than on cable-tool systems, rigs still have to "**trip out**" and take all the drilling pipe and the bit out of the hole to replace the worn-out cutting heads on a rotary bit. Casing is also still set into the hole where needed to stop the walls of the well from falling into the hole.

As with a cable-tool hole, the operation can go dreadfully wrong, or "**south**," as a crooked hole, oil or gas blowouts, and many other accidents can slow down or defeat the driller. Drilling became more complicated and dangerous over the years, so in 1949 the Alberta government and several industry associations created the Petroleum Industry Training Service (PITS). In 2005, PITS amalgamated with the Canadian Petroleum Safety Council into a new company called Enform; this organization offers training, certification, and health and safety programs for the upstream industry. It trains workers for jobs on rigs, deals with safety issues, and teaches them how to deal with oil spills and meet environmental regulations, as well as many other aspects of life on a rig.

Drillers are now making crooked holes on purpose. In the early days a hole that went off its chosen path was a disaster, but modern companies have invented complex directional drilling programs. Horizontal drilling, as directional drilling is sometimes called, can help maximize production from a formation. Once the well hits the production zone, the drillers change its direction to follow the layer of the rock that contains the oil and gas. Instead of drilling many wells to puncture the petroleum seam, one well can do it all.

Most wells do not flow on their own. In 2000, for example, Canada had more than 53,000 oil wells in production, but only 5,088 of them were flowing freely. The rest needed help—**artificial lift**, it's called—to suck the oil out of the ground. Pump jacks are one way to get oil out of a stingy field. Water or pressurized gas can help force oil out of the rock too, but drilling more wells, and in new directions, is often necessary.

Most oil wells do not flow freely—they need help.

Though used on land too, directional drilling is often used from offshore drilling platforms. In some cases, massive drill ships hover over the ocean floor as they drill. Drilling platforms also serve as a base for drilling. They can float or be made so their base sinks down to the ocean floor, in relatively shallow locations, anchoring the whole operation securely. But icebergs and storms can threaten even these large drilling installations, so some companies build artificial islands or drill through permanent ice where conditions warrant.

From shovels to directional drilling, drillers have used many tools and innovative forms of technology to solve the challenges of getting petroleum out of the ground and into the pipeline to our homes, cars, and factories. ∎

Key Dates
Numbers of Cars in Canada

1898 –	First car came to Canada
1905 –	565 cars
1955 –	2,960,900 cars
1971 –	6,967,247 cars
1986 –	15 million vehicles
2007 –	20 million vehicles

Consumption of Oil

Average person in India –	half a barrel per year
Average person in developed nations –	12 barrels per year
Average Canadian –	25 barrels per year
Average Saskatchewanian –	50 barrels per year
Average Albertan –	60 barrels per year

Consumers
Using Twenty-five
Barrels of Oil

Afamily of four people uses 100 barrels of oil each year. That's a stack of oil drums the width of a normal city lot and six rows high, so high you can't see over the wall of metal drums.

If you take the total consumption of petroleum in Canada and divide it by the number of Canadians, you get 25 barrels per person. On average, we each use that much petroleum for heating, in our cars, for generating electricity, in manufacturing, and in many other ways.

A barrel is 159 litres, or 42 U.S. gallons, or 35 Imperial gallons. If you're an average Canadian, you use about 4,000 litres (1,057 gallons) of oil products each year, and more if you live in Saskatchewan or Alberta. That's a lot of petroleum! How do we use so much, and why do the people who live on the prairies gobble up so much more than the national average?

From Kerosene Lamps
to Henry Ford

In 1900 most Canadian families relied on some sort of oil for light in their homes, consuming about half a barrel per family each year. That was all, and it is almost nothing compared to the many ways we use oil today. They used none in cars or airplanes, of course, nor did they use oil in manufacturing. When British American inaugurated its first refinery at Toronto in 1908, it produced mostly kerosene for lamps.

The first car arrived in Canada in 1898. By 1905 there were 565 of them. Imperial Oil opened the first service station at its

Service station in Calgary, no date. (Glenbow Archives, NA-2948-3)

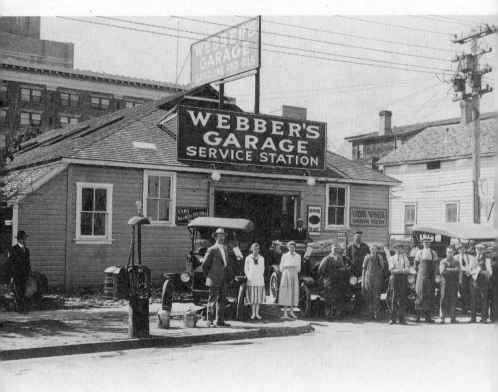

Vancouver warehouse in 1908. A garden hose connected to a kitchen water tank filled with gasoline served as the first gasoline pump. Elsewhere, clerks stored gasoline in cans behind the counter or in a closet, and you had to ask for it, like a prescription drug. Almost no one in Canada had a car.

That all changed after World War 1 when the Good Roads movement began, calling on the government to build paved roads. Founded by bicyclists who wanted to be able to ride safely in all weather (mud and gravel roads can be dangerous when wet), the Good Roads movement really took off after cars became popular.

Until 1917 the materials necessary for paving roads were all imported, but with the construction of a special refinery in Montreal East, Canada had a plant to support the construction of paved roads. There were also six other refineries in Canada of "commercial importance." There were two in Vancouver and one each in Toronto, Petrolia, Sarnia, and Regina. These refineries handled products from Canadian oilfields, yet also had to import petroleum from Illinois, Oklahoma, California, and Wyoming, as well as Peru, in order to meet the Canadian demand for oil products.

By 1920 petroleum and its derivatives were used for light, heat, and lubrication, to fuel **internal combustion** engines, to pave roads, and in many other applications. It was cheaper and superior to animal fats, such as whale oil, one of the products it replaced in domestic consumption. Economical and relatively clean oil stoves replaced their coal and wood ancestors. By 1917 a factory in Sarnia, Ontario, was producing oil stoves and heaters for Canadian consumers at a rate of 100,000 each year.

But it was Henry Ford's plain black automobile that really got us hooked on petroleum. The internal combustion engine was a revolution. Thousands of little explosions of gasoline in small chambers called cylinders gave us the power to drive vehicles all over the continent. In 1914 there were more than 50,000 vehicles on the road, or one for every 280 Canadians.

The two World Wars made good use of self-propelled vehicles like cars and trucks, and weapons like tanks and airplanes. Diesel engines became common in heavy industrial use in ships, large trucks, tanks, tractors, and railway locomotives. From the 1920s to the 1940s industry went through a transition period from coal to diesel, and for a time ships and trains that had used coal retrofitted their boilers to use fuel oil. By the 1950s a low-grade petroleum, like kerosene, was in common use as a fuel in the jet engines of a fast new breed of airplanes.

C. F. Smith, Mrs. C. F. Smith, Elmsley Smith, and Marian Smith in early Model T Ford in Medicine Hat, Alberta, 1905. (Glenbow Archives, NA-1824-1)

The Switch from King Coal to Prince Petroleum

The year 1951 marked a milestone in the history of Canadian petroleum. Although Canada had steadily come to rely more and more on oil and gas during and after World War II, in 1951 the value of Canadian oil and gas production, at about $124 million, exceeded that of coal, at just under $110 million. Coal was no longer the king of the industrial world. For the record, though oil and gas replaced coal in many transportation systems, coal is still used extensively to make electricity and in factories that create steel and other products. Wood also became less common in industrial use, but consumers still burn it in fireplaces, wood stoves, and when we go camping.

About 2,960,900 passenger cars motored around the few roads by 1955, over three times the number registered in 1928. People were switching over to petroleum for nearly every part of life. More than 40 refineries were in use by the 1950s. In 1928 Canadians had used 20,454,000 barrels of oil, or about 2 barrels per person; by 1955 each Canadian was using almost 13 barrels of oil a year, for a total of 201,285,000 barrels.

For many generations, Canadians imported oil from the United States, South America, and the Middle East. We only became self-sufficient as a result of discoveries in Alberta in the late 1940s and through the 1950s. As of December 31, 1957, Canada's production potential of 900,000 barrels per day exceeded its consumption of 750,000 barrels per day, making the country technically self-sufficient in petroleum. However, the Americans lifted import restrictions in the later half of that same year, after the Suez Crisis was resolved (see Bitumen—The Oil Sands, pp. 99–119, for more details on the Suez Crisis). In an attempt to retain the markets it had gained during the crisis, the United States also flooded Canadian markets with its own product, threatening Canadian production.

Natural gas is another important petroleum product, but in the early days most of it was flared during the production of liquid gasoline. A few small gas fields provided fuel to local markets in the late 1800s, and in 1912 a 270-kilometre- (168-mile-) long pipeline connected the Bow Island gas field in southern Alberta to Calgary. Edmonton got connected to gas in 1923 via a 130-kilometre- (81-mile-) long pipeline from the Viking field in northeastern Alberta. During the 1950s and 1960s more gas was discovered along with the oilfields throughout Alberta and other places in the West. These discoveries allowed the construction of the first of many gas pipelines to central Canada, the coast of British Columbia, and to Washington State and California. Propane, butane, and other gases that had been burned as waste byproducts found uses, too. For example, propane companies began marketing it as an alternative to natural gas, one that could be bottled and delivered to remote rural consumers who had no access to a natural gas pipeline. Butane, pentane, and other gases found use in consumer appliances like stoves and furnaces and were used to make aviation gasoline for high-performance airplanes. Gas-processing plants added recovery units to their operations in the early 1950s to make sulphur, which was used for rubber, fertilizers, and other manufacturing processes.

As a clean-burning alternative to coal and oil, natural gas became the fuel of choice for many applications. Gas discoveries in the North since the early 1960s prompted the development of a natural gas pipeline along the Mackenzie Valley. Though it was delayed in the late 1970s, the current high prices and demand for petroleum products of all types promise that a pipeline will one day bring large quantities of petroleum out of the distant reaches of the Canadian North.

Canadians changed their transportation habits in the decades after the war. The number of passenger cars jumped from 2,960,900 in 1955 to 6,967,247 in 1971. That amounted to a change from one car for every five people to one for every three people—at least one per household. In 1971 the total number of vehicles, including trucks, buses, and delivery vehicles, was 9,022,136. The lust for cars in Canada paralleled their demand in the United States and Europe. Although some were made in Canada, most were imported. An analysis of the situation in the *Canadian Annual Review* for 1960 stated:

> No industry in Canada reflected the effect of the consumer demand for imported rather than domestic products as clearly as the automobile industry. Combined with this trend was the parallel consumer demand for the new "compact" models. During 1960 about 32 percent of total car sales in Canada was British and European imported models, while the compacts accounted for about 17 percent and the remainder consisted of the American-style "big" cars manufactured by the Canadian subsidiaries of the US automobile manufacturers.

When Prime Minister Diefenbaker officially opened the Trans-Canada Highway through the Rogers Pass on September 3, 1962, the road link across Canada was complete. Our love affair with the car continued, and by 1986 almost 15 million cars, trucks, and buses were being driven by 25 million Canadians. As of 2007 our population of 33 million people used 20 million vehicles—that's almost two vehicles for every three people. With all these cars overtaxing the capacity of our roads, is it any wonder that many Canadians spend more than an hour a day commuting to and from work?

Heavy Users

We do more than drive cars. In fact, at 25 barrels per capita Canadians are some of the heaviest users of petroleum in the world, and that number rises to about 60 barrels in Alberta. Feeling guilty? Read on further before you decide to move to India.

First, we use lots of oil and gas *because* we produce it in great quantities. Canada has been self-sufficient in petroleum products for several decades, and the country exports more than twice as much as it imports.

Road-building crew working on Trans-Canada Highway, Alberta, 1957. (Glenbow Archives, ND-10-227)

What They Do...
Petroleum Engineers: Coaxing Oil Out of Rocks

Petroleum engineers get the oil out of the ground. Oilfields produce oil, gas, water, sulphur, and lots of other materials, so it's up to the engineers to figure out a way to make the oil wells work. These scientists take university degrees in engineering that provide them with extensive knowledge of drilling techniques, how **reservoirs** work, and how fluids move through rock, sand, and soil.

Computers are important to the drilling, production, and reservoir engineers who work in the oil patch and help them manage large amounts of information. They work closely with **well-log analysts**, specialists who use sophisticated tools to assess the chances that oil-bearing rock in that region will give up its treasure.

Drilling engineers make sure the multi-million-dollar well has a chance of finding the oil that production engineers bring to the surface. Finally, reservoir engineers maximize the potential of the entire field.

When there are no oil well blowouts or pipeline breaks, you know the engineers are doing their job well. When their activities hit the news, they're having a bad day! ■

Each individual in Alberta does not consume more than twice the national average. The difference isn't because of pickup trucks and SUVs; it's the Alberta economy as a whole. It takes a lot of energy to search for oil and gas, to produce, refine, and process it, and then push it down the pipeline to markets all over North America.

Natural gas, for example, has to be processed to remove liquids, water, and hazardous products like hydrogen sulphide, a deadly toxin that can paralyze the lungs and kill a person, even in very small quantities. Gas plants remove hydrogen sulphide and many other impurities, including water, and make natural gas safe for use in the furnace in your home. It is also the **feedstock**, or raw material, for many other products, including antifreeze, plastics, and other items we use every day.

Natural gas is also a preferred method for making electricity in many parts of North America. It burns cleaner than fuel oil, which is why we use it in furnaces, and it is easy to move through pipelines. It can be stored in underground reservoirs, like caves, and then used later to generate electricity when demand for power is high. **Cogeneration stations** that generate electricity also make heat, which is sometimes used in nearby greenhouses or other industrial facilities.

The **oil sands** operations also use enormous quantities of natural gas and water to separate the oil from the sticky bitumen. During the 1990s and the early part of the twenty-first century, Alberta expanded its use of natural gas more than 60 per cent. On average, it takes between 750 and 1,500 cubic feet of natural gas (enough to heat a house for a week) and up to five barrels of fresh water to process and upgrade just one barrel of oil. Gas is also used to heat water to make steam, which is injected deep underground to heat the oil sands *in situ*—in their original place. After it percolates through the bitumen, the steam encourages the oil to seep out; then oil wells pull the **slurry** to the surface and send it off through a pipeline for processing.

These many ways of using natural gas are a concern. Each day the oil sands operations use enough natural gas to heat 3.2 million homes for that day. If Alberta is to increase its production of petroleum from the oil sands, producers will need to streamline this process and become more efficient. We can't use all our natural gas—and water—to suck the petroleum out of the oil sands! (See Bitumen—The Oil Sands, pp. 99-119, for more details on the resources required to extract crude oil from the oil sands.)

Oil also requires a lot of work before it can be put into our cars.

It has to be refined, a process that takes heat. **Heavy oil** and bitumen—the molasses-thick stuff from the oil sands in Alberta—are far too thick to put into a refinery, so they have to be upgraded, or mixed with lighter oil, before they can run through the complicated machinery that creates the oil and gasoline we use in our cars.

Overall, the individual Alberta consumer uses no more oil and gas than the average Canadian; it's just that it takes oil to make oil. But we Canadians still use a lot of petroleum products, and knowing how we use them can help us make wise choices every day.

Depending on where you live, you could be using more fuel to heat your home than to run your car. Canadians live in one of the coldest countries in the world. Some of us use natural gas for heat, but in central and eastern Canada we also use heating oil to keep our homes comfortable in the winter. Most homes in tropical countries don't have air conditioning, but almost every Canadian house requires heat in the winter to keep people alive and prevent water pipes from freezing. While other countries rely heavily on coal or nuclear energy to provide space heating, almost all Canadians rely on petroleum products to stay warm.

Despite our heavy consumption of fuel for heat, cars remain the largest single cause of petroleum use by consumers in Canada. Back in the 1970s, the Arab-Israeli war in the Middle East threatened the flow of oil from that petroleum-rich region to the United States and other industrialized countries. As a result, the United States government forced North American car and truck companies to produce more energy-efficient vehicles—for a while. They reached maximum efficiency in the 1980s but have been guzzling more gasoline ever since we started buying larger cars and SUVs. Cars in Europe are smaller and at least one-third more efficient. Gasoline prices in Europe are also much higher, they don't have as many vehicles, and they don't have to drive them as far as we do here.

What They Do...
Pipeline Crews: Long Ribbons of Steel

Pipeline workers make sure oil and gas get to your house and car. After the pipeline is laid in a trench in the ground, dedicated men and women make sure it operates efficiently. These skilled people have high school diplomas or university degrees and work well with computers. They keep an eye on many detailed monitoring systems, preventing accidents and explosions along the line.

In the early days the trenches were dug by hand, and the small-diameter lengths of pipe were screwed together, like a water hose. Today's pipeline workers use large trenching machines and automated welding robots to install large-diameter lines. Though pipelines are nearly invisible, tens of thousands of kilometres of them criss-cross Canada and the United States, and the people who keep them working smoothly are important to our economy. ■

Lowering the speed limits on major highways throughout North America helped reduce gasoline consumption in the 1970s and 1980s, too. Travelling at 90 kilometres (56 miles) per hour instead of 110 kilometres (68 miles) per hour can save 20 per cent of the gas in the tank of your car. The economic downturn of the 1980s reinforced these good ideas with a hit in the wallet. Unfortunately, as governments relaxed regulations on car companies and the economy rebounded, the price of gasoline at the pump fell, and many of us returned to our previous habits. Consumption grew as speed limits rose again, and large cars and trucks became common. By using technological advantages such as fuel injection, aluminum engines, and turbo chargers, car and truck companies were able to put large vehicles back on the road and market them as "even more environmentally friendly," although they were just using tricky advertising slogans to help us feel good about wasteful consumption.

A strong economy can actually work at cross-purposes with conservation and common sense. As more people move into cities, urban areas expand and require more roads, highways, and expensive transportation systems. Housing in new subdivisions or on parcels of land outside the city becomes more attractive, especially to people who can afford large homes and don't mind driving long distances to work.

But there are other options. Commuting by public transit—common in larger cities in the world—is becoming more accepted in Canada. Buses, trains, and subways are part of every transportation system in bigger Canadian cities. Walking or biking is possible for some of us who live in cities. Choosing smaller, more energy-efficient vehicles, as well as hybrids and electric vehicles, are other ways we can reduce our consumption. It's true that even bigger SUVs have made a comeback, but small three-cylinder vehicles like the diesel SmartCar have attracted the attention of Canadian consumers too.

Twenty-five Barrels or Ten—
The Choice Is Ours

Can we continue to use petroleum resources at our current rate indefinitely? Probably, but the cost will rise, and the work required to find 25 barrels of oil for each Canadian will become even more complicated.

For example, Canadian petroleum production from all sources only increased slightly between 1998 and 2005, but it took almost three times as many oil and gas wells to produce the same amount of oil in 2005 as seven years before.

There was a time when it seemed we had an almost unlimited supply of oil and gas in Canadian reserves. When we first started exporting natural gas we did so only after making sure we had enough to supply our own needs for twenty-five or thirty years. Of course the potential for finding oil and gas exists, and there are known reserves of petroleum from the oil sands for decades to come, but the *proven* reserves of **conventional oil**—the cheap stuff to produce—are only enough to assure our current rate of production and consumption for nine years, and the same holds true for natural gas. That's why companies are pushing for more production from the frontier regions in the North and offshore, as well as from non-traditional sources, like methane from coal beds, and from hard-to-reach drilling locations in mountains and areas far from pipelines. They are even looking at expensive systems for liquefying natural gas in Russia and moving it by tanker to consumers in eastern Canada.

Consumption will continue to rise if we do nothing to curtail our current habits. The price of petroleum products will rise. Reaction to economic forces will force us to change—and we have changed many times in the past.

Perhaps the best indicator of the future of oil lies in the advertisements placed in magazines by the big oil companies. They talk about investing in wind power, in solar research, about making fuel from corn and other plants. They challenge us to join them in the creative process of reforming our habits, individually and as a society. One oil company in downtown Calgary relies on solar collectors to help heat and cool its office tower. Maybe we can learn from them how we can use less than 25 barrels per year. ∎

Key Dates

Macdonald National Policy –	1870
Laurier Petroleum Bounty Act –	1904
Mackenzie King and Wartime Oils –	1943
Diefenbaker National Oil Policy –	1960
Lougheed Natural Resources Revenue Plan –	1973
Clark and the Energy Self-Sufficiency Tax –	1979
Trudeau National Energy Program –	1980
Mulroney Western Accord and Atlantic Accord –	1985

Conflict
Seven Examples of Canadian National Energy Programs

Just imagine, if you can, that 72 percent of the revenues in the oil and gas industry in the United States went to foreign-owned companies in 1979. Or imagine that 82 percent of those revenues went to foreign-controlled companies. Given that situation in the U.S., do you suppose it might have become an issue in your recent election campaign?

—Marc Lalonde, federal energy minister, in defence of the National Energy Program in New York City, November 19, 1980

The **National Energy Program (NEP)** of 1980 was a disaster—everybody says so. The 1979 version would have saved the world—just ask the planners of the day. The difference is that the NEP of 1979 died when Prime Minister Joe Clark's government had to call an election, and the Pierre Trudeau 1980 version was implemented just as the world oil price peaked, interest rates took a big hike, and the global economy took a nosedive for several years.

Programs, policies, and plans are all bad words. Central governments always get into big trouble when they create a national strategy for anything, whether it is child care, guns, wheat, booze, railways, water, or oil.

The role of each federal politician is to do what is best for the individuals in the constituency and for the country as a whole. Elected members of Parliament from each region of the country gather in Ottawa to conduct the nation's business in a way that works the best for everyone, one hopes. That's democracy, but democracy is also the main reason for so much misunderstanding and tension across this big country. Attempts to develop oil resources in Canada date back to the country's earliest days, and the political strife oil policy has created is rooted in Canadian history.

The Macdonald National Policy

The term "national policy" is older than Canada and refers to a centralized economic strategy, designed to encourage economic growth. The Company of Adventurers of England Trading into Hudson's Bay was formed in 1670 by Charles II of England and given a charter for "the sole Trade and Commerce" in all lands whose waters drained into Hudson Bay. The commerce in the region was the fur trade. Never mind that the land did not belong to the King of England and that the people already living in that part of North America were not consulted. The right to do commerce in most of Canada was granted to a small group of businessmen by a foreign government.

The same thing happened with all natural resources, including petroleum. Even before 1867 and the formation of this country called Canada, the Geological Survey was in place. Beginning in 1842, this arm of government began looking for minerals so that businessmen could develop them and strengthen the economy of the Province of Canada—a small part of what is today's Ontario and Quebec. The economy needed minerals if it was to prosper, especially coal, which was central to the steam engines that powered industry and railways.

Sir John A. Macdonald promised to develop the West for the good of the new country.

On July 1, 1867, Canada became a country. The act of Confederation joined Upper and Lower Canada with Nova Scotia and New Brunswick into a large, but tentative, country. British Columbia signed on in 1871 on the promise of a rail link to the rest of the country. In order to keep this promise, Prime Minister Sir John A. Macdonald devised a three-part plan to develop the Canadian West for the good of the new country:

Geologists Ken Huff and O. A. Erdman at head of Rocky Mountain Canyon on the Peace River, Alberta, 1942. (Glenbow Archives, PA-2166-116)

first, bring in lots of immigrants; second, charge high tariffs to protect Canadian industry from foreign competition; and, finally, build a transcontinental railway to link the areas with natural resources, such as coal mines, to the factories in central Canada. Macdonald called it the **National Policy**, and it was the first of many such policies. It was controversial because it exploited the natural resources of one region for the benefit of another part of the country.

Macdonald's National Policy was built on coal, not petroleum, but fuel is fuel. Railways and factories needed coal to operate, and so explorers went out across the country looking for the hard black gold to burn in steam engines. Along the way they noticed oil and gas too, but with no markets for these products they just kept looking for coal. They found it in the mountains in the West, first in Alberta at Canmore and the Crowsnest Pass, and later in British Columbia. Coal fired the railways of Canada for almost a century before diesel engines began pulling trains across the country in the 1950s.

Canadians had discovered oil in southwestern Ontario in the 1850s, and by 1863 there were about thirty refineries in Ontario. But Canadian oil smelled bad. Oil and gas that contain sulphur (hydrogen sulphide) are called "sour" due to the smell, and need to be cleaned, or "scrubbed," in order to make them "sweet." Canadian oil smelled like rotten eggs, so most consumers preferred oil from the Pennsylvania oilfields across the border in the United States. In 1867 the government of Canada imposed a tariff on imported oil to protect Canadian markets from competition. Sir John A. Macdonald was the first prime minister to meddle in the Canadian petroleum industry.

Canada has been a net importer of oil for most of its history.

Though we began exporting oil in 1868, Canada has been a net importer of oil for 95 of the past 140 years. That is, it imported more than it exported. It was self-sufficient in oil from 1881 to 1885, from 1959 to 1973, and has been a net exporter again since 1984. In 1880 sixteen Canadian companies joined forces to fight American competition and created the Imperial Oil Company. It lasted as an independent Canadian company for eighteen years until it was gobbled up by Standard Oil, an American company, in 1898—so much for Canadian control over its oil industry.

Turner Valley oilfield and village, 1942.
(Glenbow Archives, PA-2297-111)

The Laurier Petroleum Bounty Act

Prime Minister Wilfrid Laurier's oil policy tried a new angle. Beginning in 1904, Ottawa paid a subsidy of 1.5 cents for every gallon of oil produced in this country. The **Petroleum Bounty Act** of 1904 expanded the program to include petroleum extracted from oil shales. On July 1, 1924, the subsidy fell to 75 cents, and the program ended on July 1, 1925. During these twenty years the Department of Trade and Commerce paid out more than $3 million in bounty to producers of oil.

*Petroleum bounties and subsidies
are a Canadian tradition.*

Mackenzie King's Wartime Oils

William Lyon Mackenzie King's oil policies during World War II affected the Alberta oil patch directly. Under wartime conditions the demand for petroleum products grew quickly. Turner Valley oilfield production in Alberta had peaked in 1942, yet Ottawa was under pressure from the United States to provide more western Canadian oil for the war effort. Washington forwarded a paper emphasizing its point: "The Need for a Concerted Program of Wild Catting for Oil in Western Canada." This 1943 document encouraged George Cottrelle, the Dominion Oil Controller, to establish a Crown corporation to help find oil. **Wartime Oils Limited** began operations in 1943 with high-profile westerners on the board of directors, and with Dr. G. S. Hume of the Geological Survey of Canada as an adviser. Terms were reasonable: the loans only had to be repaid if the wells struck oil.

Although other areas were considered, Turner Valley wildcatting seemed most promising. During the remaining years of the war, Wartime Oils loans funded the drilling of twenty-two wells in the area. Twenty-one found oil. They produced more than a million barrels of oil by the end of 1946 and demonstrated Ottawa's interest in the development of western Canada's only major oilfield during the war.

Beginning in 1943, the minister of national finance allowed some drilling and exploration expenses to be deductible from taxable income. Ottawa expanded the deductions to include all exploration and developments costs in 1948, with the exception of the costs of rights and purchased properties, which it included in expanded policy that took effect in 1962. Though bonus payments were not included in these incentives, there was an indefinite carry-forward on all the remaining deductions,

What They Do…
Politicians: Struggling for Power and Control

The premiers and the prime minister of Canada control the oil industry. The international economy plays a large part in the Canadian oil industry because oil and gas are traded as global commodities, but politicians play a big role in how the Canadian petroleum industry operates.

Unlike other countries where oil and gas are owned by individuals, most petroleum in Canada is owned by the Crown—the government. That's why provincial and federal politicians fight over the way it's developed, how much it's taxed, who gets to control how fast it comes out of the ground, and where we sell it.

Power and control are important parts of this game. Almost every level of government gets involved through regulations, taxation, fees, and licences, as well as through direct subsidies to the development process and indirect support by building the infrastructure necessary to make the system work—roads, schools, hospitals, even golf courses. ∎

allowing companies to use these incentives to help finance their operations over a period of years. By the late 1960s, these deductions also included geological and geophysical expenses, buildings, and equipment, as well as payments made for exploration rights.

Diefenbaker's National Oil Policy

Next came John Diefenbaker's **National Oil Policy** in 1960. Discoveries of massive oil and gas fields in Alberta in the 1940s and 1950s had flooded the western Canadian market with product, but it was not cheap. Compared to imports from South America and the Middle East, Canadian oil was expensive, so Ottawa gave western oil a protected market—from the Ottawa Valley west—by imposing an import tariff in 1965 on cheap gasoline from Europe. It also encouraged Canadian companies to build pipelines across the international border and sell oil to the United States. By 1966 more than a million barrels of oil per day were leaving Alberta by pipeline for markets in central Canada and the United States. These programs worked well and allowed the western Canadian oil industry to grow.

Offshore companies operating in Canada did not like the Canadian policy because it prevented them from importing Venezuelan oil into key Canadian markets. (In 1962 about 70 per cent of Canadian production was owned or operated by non-Canadians, mostly Americans.) The companies convinced President Richard Nixon to impose import quotas on Canadian oil in 1970, but the United States had been a net importer of oil since 1948 and needed all the oil it could get, so Canadian exports grew, and by 1971 almost a million barrels of oil a day were crossing the border to American consumers. When the Middle East oil crisis hit in 1973 the Americans lifted their import restrictions, but by that time Ottawa had determined that Canada was facing a shortage and refused to supply any additional oil or gas for export.

"The real dispute is over power and control."
——Jim Gray, an owner of Canadian Hunter Exploration

Industry did not agree. In 1971 the Canadian Petroleum Association claimed that, at 1970 production rates, Canada had 923 years of oil reserves and 392 years of natural gas. Imperial Oil's 1972 annual report stated that reserves of petroleum were adequate to supply markets for "several hundreds of years." It warned that any lost markets due to restrictions on exports could prove impossible to regain, and could create "a genuine economic setback for Canada."

"Once a barrel of oil goes down
the pipeline it's gone forever."
——Peter Lougheed

Lougheed and Trudeau:
The Natural Resources Revenue Plan
and the Alberta Energy Company

The Blue-Eyed Sheik became premier of Alberta as the international price of oil soared from $2.50 per barrel in 1970 to $10.50 in 1973. The new premier of Alberta, Peter Lougheed, and the new Progressive Conservative Party of Alberta took control over the Alberta political scene and acted quickly to cash in on the rising price of oil. The Lougheed government told the petroleum industry that it wanted a larger share of the economic rent when it announced its intent to raise **royalty** payments as part of its **Natural Resources Revenue Plan**. It unilaterally repealed the ceiling on royalties payable by producers and created a rate that was linked to the international price of oil in 1973. In effect, the Lougheed government more than doubled its income from the oil industry. Alberta's unprecedented action was significant because it changed the relationship between the foreign-controlled industry and the provincial government in Alberta. In 2006 Lougheed said to the people of Alberta, who were once again caught in the middle of a boom, "Think like an owner … What is a fair return? We knew it would fluctuate over time … Once a barrel of oil goes down the pipeline it's gone forever. It's like a farmer selling off his topsoil."

Alberta premier Peter Lougheed spits a bit of the wealth in the direction of Prime Minister Pierre Trudeau in an editorial cartoon by Tom Innes published in the *Calgary Herald*, February 24, 1979. (Glenbow Archives, M-8000-372)

Lougheed's brother Don worked for Imperial Oil, so Peter chose his words carefully and never threatened to "nationalize" the oil industry or set up a government oil company. But hiding behind the capitalist mantra of market forces and competition, he increased Alberta's control over its oil patch by taking a larger slice of the revenue pie. In 1973 he created the Alberta Energy Company, his own "national" oil company, with the government owning 49 per cent and Alberta investors 51 per cent. That same year the Trudeau government in Ottawa announced its plan to create a national oil company, Petro-Canada, as part of a plan to increase Canadian ownership of the petroleum industry.

The Northern Pipeline

Ottawa's plan to encourage development of oil from the frontiers hit a major hurdle in the North in the mid-1970s. Though companies had made promising discoveries in the 1960s and 1970s, they relied on a pipeline to move petroleum south. The Mackenzie Valley Pipeline Inquiry began its investigation into the development of a pipeline out of the North on October 27, 1975. Hearings throughout the region gave the committee input into the effects the line would have on the northern economy, politics, and culture. In early 1977, Panarctic Oils estimated Arctic Island reserves at 16 trillion cubic feet of recoverable gas reserves. But on May 4, 1977, the Berger Inquiry into the Mackenzie Valley Pipeline released the first part of its report. It called for a ten-year moratorium on Mackenzie Valley pipeline development. Within months, Gulf Oil Canada and Mobil Oil Canada ceased drilling in the Mackenzie Delta area of the Arctic until the pipeline debate was resolved. This turned out to be for the best, because the expensive pipeline would have completed construction in the middle of the 1980s when the price for natural gas hit a record low. ■

The rush for spoils was not pretty. Premier Lougheed and Prime Minister Trudeau only got away with grabbing a larger piece of the resource pie because the price of oil escalated quickly during the next few years, reaching a peak of $44.66 in 1980. At the time, predictions called for the price of oil to go to $100 per barrel. Though the oil companies complained about the grab, they also made increased profits during this period, and Canadian oilfields were more politically stable than those in countries that were part of the **Organization of Petroleum Exporting Countries (OPEC)**. The Lougheed and Trudeau governments fought many times during the 1970s over the Canadian price of oil and the sharing of the **taxes**, trying to control and set policy for the oil and gas industry.

Clark: The Energy Self-Sufficiency Tax and the Canadian Energy Bank

In 1979, Joe Clark became the first prime minister from Alberta since R. B. Bennett in the 1930s. Westerners had high hopes for the man from High River. When challenged by the Liberals to do something to keep the foreigners from taking excess profits as the price of oil continued to rise, he said, "We are now actively discussing the mechanism that will be used to ensure that virtually 100 percent of the revenues that would go to the companies as a result of the increases in energy prices will, in fact, be regained by the Government of Canada for the specific application by the Government of Canada for national energy purposes."

"We will not achieve energy self-sufficiency by 1990 if we continue to treat energy as a cheap product that we can afford to waste."
——Joe Clark, 1979

Prime Minister Clark was wrestling with Alberta too, just as Trudeau had throughout most of the 1970s. Peter Lougheed was a worthy opponent, not one to back down easily from his plan to make sure that Alberta benefited from the temporary oil boom. But the Tory government in Ottawa had the power to make national policy, and it promised to continue holding the price paid for Canadian oil to below the world price. It wanted to impose an "energy self-sufficiency tax" worth $2 billion a year that would split any big windfalls evenly between the producing provinces and Ottawa. Ottawa's take from the oil patch would have increased from 10 to 19 per cent of the overall revenue, while the industry's share would have dropped from 45 to 37 per cent and the provinces' portion would have remained stable. Ottawa also intended to grant new export licences for natural gas to the United States and use the expected income from these various sources to encourage additional development of Canadian reserves. Clark's proposal also included a **"Canadian energy bank"** to invest in projects that were important to Ottawa—especially ones by Canadian-owned companies.

Clark's plans were as Canadian as maple syrup and *Hockey Night in Canada*, but the real kicker was the tax at the gas pump. The Conservatives fought the next election in early 1980 on the basis of an aggressive national oil policy that included a 25-cent per gallon tax at the pump. Gas prices went up 18 cents overnight, or 257 per cent. The vastly increased tax was a direct attempt to force reluctant Canadians to "voluntarily" accept conservation measures and assist in the Conservatives' goal of making Canada self-sufficient by reducing its reliance on imported oil (we had been a net importer since 1974).

The Conservatives lost the election. In spite of their ambitious attempts to bring harmony and fairness to the contentious regional debate over petroleum prices and taxation, the Clark government found itself unable to balance the interests of the producing and consuming provinces. Resource security, income sharing, and control over the development process proved no easier for the Conservatives to juggle than the Liberals.

Trudeau's National Energy Program

The new Liberal government in 1980 found Alberta no more willing to negotiate than during the 1970s. So on October 28, 1980, it announced its National Energy Program (NEP) as part of its new budget. It promised a Canadian oil price, at least 50 per cent Canadian ownership of the industry by 1990, and self-sufficiency in oil by 1990. It basically sought to insulate Canada from volatile international oil prices. The goals were commendable.

"No more mister Nice Guy" says Alberta premier Peter Lougheed as he enters a bar where Prime Minister Joe Clark, Liberal leader Pierre Trudeau, and Ontario premier Bill Davis are discussing ways to divide up Alberta's oil wealth. A cartoon by Tom Innes published in the *Calgary Herald* on November 1, 1979. (Glenbow Archives, M-8000-486)

Just like Joe Clark's national oil policy, the implementation of the NEP pitted region against region. It discouraged some types of investment while subsidizing others. To put it mildly, the NEP was received poorly by industry. The negative reaction caused many companies to postpone their development plans, especially in areas of new technology. The oil companies also attacked the concept of self-sufficiency in oil as an artificial idea and unfair if no other commodity was being singled out for this kind of special treatment. Jim Gray, an owner of Canadian Hunter, a Canadian oil and gas company, wrote in 1981, "The solution to our energy problems is in a very real way tied to the solution to Canada's broader constitutional debate. The debate is being waged on an ideological front, on a political front. The real dispute is over power and control."

The NEP was Ottawa's forecast for the future and its plan for economic prosperity for Canada as the price of oil continued to rise. While unpopular with the producers, the Liberals believed it would be popular with the voters in central Canada and stabilize the economy for the predicted boom of the 1980s. In 1981 Ottawa approved the Norman Wells oil pipeline but delayed it eighteen months in order to settle Native land claims. Later that year, Petro-Canada takeovers began as part of the NEP program of attempting to make the industry at least 50 per cent Canadian. Its first purchase was 71.5 per cent of Petrofina Canada's assets.

The events of 1980 help us understand the nature of the relationship between the federal and provincial governments during this crucial period. As a result of a set of programs endorsed by both major political parties—the Conservatives and the Liberals—and implemented by the Liberals in 1980, Canadians gained an increased role in the development and operation of their petroleum industry.

For better or worse, the NEP of 1980 proved that the federal government had the power to effect change in the face of over-whelmingly powerful international political events and multinational economic forces, to say nothing of the regional interests within Canada.

Public response to the NEP was very positive in most of Canada, and a Canadian Petroleum Association poll showed 84 per cent support by Canadians for the goal of making the petro-leum industry at least 50 per cent Canadian. Canadians also overwhelmingly supported the expansion of Petro-Canada to make it Canada's largest producer, even if that meant buying up Imperial Oil, Shell, Gulf, or Texaco.

In the early 1980s the world price of oil began to fall from almost $45 per barrel to a low of $19 in 1988. As the price of oil dropped, oil issues became less violent, and an agreement between the federal government and the producing provinces allowed the Canadian price to rise to almost world oil levels. Prices had been kept below world levels in order to control inflation in oil-consuming provinces. Megaprojects stalled due to the instability of the economy, the recession, and high interest rates that peaked at 21 per cent in 1981.

Although things slowed down for the industry, provinces

Revenue from Canadianization Programs

Canadian-controlled companies earned 28 per cent of all petroleum revenues in the country in 1983, up from just 13 per cent in 1977 and 19 per cent in 1980. As a result of **Canadianization** programs, that figure rose to 48 per cent in 1986. Despite a different direction taken by Ottawa after the Brian Mulroney government took control in 1984, Canadian-controlled companies still generated more than 46 per cent of the revenues in 1994. ∎

Why Westerners Hate the NEP

Westerners have always felt that the East—central Canada—is ripping them off. The 1980 NEP felt like just another example.

In the middle of the biggest oil boom to date, the NEP forced new restrictions on the West, gave Ottawa more control over the oil industry, and took billions of dollars out of the economy in new taxes. Provincial governments howled, and oil companies threatened to leave. Westerners felt alienated and betrayed yet again.

Then, as the economy fell apart, interest rates skyrocketed, people lost their jobs, their homes, and their savings. Two Alberta trust companies went bankrupt too. Drilling rigs left Alberta, headed south. The price of oil fell, and so did the income to the Alberta taxpayer. Even the Alberta government went into debt.

The worst recession since the 1930s hit the whole world, hard. The NEP did not cause the worldwide downturn, but it certainly made it worse in western Canada, prompting Ralph Klein, then mayor of Calgary, to utter the words that symbolized the emotions of many westerners at the time: "Let those eastern bastards freeze in the dark."

Any mention of the NEP still raises the heat under the collars of many westerners to this day. But each prime minister, regardless of political persuasion, has to keep a careful eye on an industry that is so central to the economic and political pulse of Canada. ■

like Alberta still relied heavily on the oil patch. In 1984 income from petroleum royalties, land bonuses, and land rentals totalled $5.2 billion, or more than half of the revenue for that year.

Prospects for offshore development, a frontier area that got special attention in the NEP, got better in the early 1980s. In early 1982 Ottawa and Nova Scotia signed an offshore development accord. During 1983 Newfoundland signed its first offshore exploration agreements. Ottawa's attempts to encourage the industry to exploit the frontiers began to show success just as the price of oil fell.

Mulroney: The Western Accord and the Atlantic Accord

In 1984 Brian Mulroney and the Progressive Conservatives won a landslide election and quickly began to carry out their election promise to deregulate the oil industry—a trend that had started in the United States in the 1970s—and to do away with the NEP. The Conservative government immediately began dismantling the NEP in its economic statement of November 1984. In 1985 it signed the **Western Accord** with the western provinces, an agreement that ended eleven years of controlled oil prices, relaxed oil export restrictions, abolished five federal taxes from the 1980 NEP, and phased out incentives and grants to the industry.

Also in 1985, the federal government and the government of Newfoundland and Labrador signed the **Atlantic Accord**, which stipulated that the two levels of government would share the management of offshore oil and gas resources as well as income from the exploitation process. Its provisions ensured that benefits would flow to the province and the country as a whole, guaranteed self-sufficiency and security of oil and gas supply for all Canadians, and provided a stable political and economic climate to allow industry to develop the resource.

> *"Go and find oil and gas
> and sell it on the open market."*
> *—Pat Carney, federal energy minister*

But when world oil prices fell further, government revenues also fell. Decontrolled oil prices came into effect on June 1, 1985, for the first time in twenty-five years, just as the price of oil began to plummet. The Atlantic Accord helped set the stage for the development of petroleum off the East Coast, while the Western Accord gave up revenues to Ottawa totalling almost $4 billion until 1991. The last provisions of the NEP were finally rescinded on October 31, 1985.

The new Conservative government gave the petroleum industry exactly what it wanted: a deregulated price as of July 1, 1985. Just as the price plummeted to little more than half the 1980 price, it hit $22.83 in 1986. The new energy minister, Pat Carney, issued a challenge to the petroleum industry after signing the Western Accord on April 3, 1985:

> We delivered, now it's industry's turn. With dereg-
> ulation of crude oil prices and export controls
> and with the scrapping of numerous levies, taxes
> and charges, I am challenging the industry. It's a
> clear cut message—go and find oil and gas and
> sell it on the open market.

Industry claimed to be happy with the changes, but insisted the federal government **grandfather**, or continue to honour, existing petroleum incentives programs worth billions of dollars to the companies—funds worth as much as $2.5 billion or more. The associations that represented oil and gas companies welcomed the deregulation process, but argued that the various levels of government had to allow the income liberated by the demise of the taxes to flow back to the producers and not be recouped in other provincial or federal royalty and tax programs. Their pleas for further reductions in royalties met with success in April 1986 when the Alberta government announced a $400 million program of incentives for small producers. That same month representatives from four hundred small companies met in Calgary to request a federal cash flow stabilization program, thereby seeking intervention from the very level of government they had so vociferously denounced for its intrusive policies for more than a decade. By the end of the year Alberta announced another $1 billion in royalty cuts and tax credit extensions to benefit industry over the next five years.

Policy for the Twenty-first Century

During the 1990s and the early years of the new millennium, the Canadian petroleum industry continued to evolve. The older oilfields and regions have become "mature" or have reached a peak in production. New technology will perhaps allow them to continue to produce for decades, but they can no longer promise an increased supply. Frontiers offer the best hope for the future, but these areas in remote parts of Canada and offshore require expensive infrastructure. Oil and natural gas production off the East Coast came on-line in the 1990s, and drilling in the Canadian West set records as companies drilled more wells in order to replace dwindling reserves. New gas fields in the Northwest Territories added to the reserves too, and the rising price of oil during the past few years has made expensive projects economically viable.

In 2001 the production of oil from the oil sands in Alberta exceeded conventional oil production for the first time, reaching 271 million barrels compared to 264 million barrels of conventional oil. If the price of oil stays high, the reserves in the oil sands are potentially unlimited, but the economic, social, environmental, and political challenges associated with developing this resource are complicated (see Bitumen—The Oil Sands, pp. 99-119). In 2002 companies began developing **coal bed methane (CBM)** projects (see Frontiers—Oilfields on the Edge, pp. 121-35), another non-traditional source of natural gas. The record for wells drilled in Canada also continued to rise each year, with more than 28,000 in 2005.

The minority Conservative government elected in 2006 has faced many policy challenges. Alberta's outgoing Conservative premier, Ralph Klein, warned Prime Minister Stephen Harper not to make a grab for Alberta's oil wealth, and the federal government has been careful not to step on Alberta's toes. Neither side wants another NEP, but a national energy plan or strategy is needed because, if Canada does not create its own, the Americans will make decisions for our economy. For his part, Harper said he believes in "a free exchange of energy products based on a competitive market—not self-serving monopolistic political strategies." Though he was referring to the Russian economy, Harper's faith in free markets may mean he has already turned over the reins of the Canadian petroleum industry to the oil companies and the United States. For example, the **North American Free Trade Agreement (NAFTA)** prevents Canadians from limiting supplies to United States in an emergency. Under it, Canada has also quietly become the largest supplier of oil and gas to the United States. Since signing NAFTA in 1994, oil exports have gone from 44 per cent to 63 per cent of our production and natural gas exports from 41 per cent to 56 per cent, and American demand for oil increases every year. By action, if not design, the Canadian government's current national energy strategy is to expand exports to support American needs.

Both Harper and Klein promised even more production to the Americans in 2006. Most of the supply will come from the oil sands because conventional oil is in decline. With the price of oil high, profits from the oil patch are lucrative, but the Alberta government only charges a royalty of 1 per cent on oil sands operations until the companies make back all their initial investment. Though the Alberta government is wallowing in billions of dollars of income from the oil industry, its take as a royalty percentage is about as low as when Peter Lougheed became premier. He and other leaders in Alberta are calling for a royalty review and new rates so as to more fairly compensate the people of Alberta for the resource.

Royalty rates vary greatly worldwide, depending on numerous factors. For example, in Alberta the rights to develop oil and gas resources are under the control of landowners, governments (municipal, federal, provincial, or territorial), railway companies, irrigation districts, Native bands, and many other groups. Each owner can negotiate different rates and conditions. And in some places, like Alaska, each resident receives an annual petroleum dividend cheque.

The value of the resource being produced is also reflected in the royalty rate; heavy oil and bitumen from the oil sands are worth less on the international market because they require more upgrading to turn them into products for the consumer. Rates also reflect the cost of developing the resource, its distance from markets, the boom and bust cycles of the industry, and the relative political and economic stability of the region where the resource is produced.

Also, royalties are only a part of the "economic rent," or charges that are part of the resource development process. There can also be fees for the right to explore and to access land for production. Taxation at every level of the process adds to the costs. Some governments offer subsidies and grants to encourage the development of oil and gas in marginal areas for complex economic, political, and social reasons. Even international wars and other foreign policy initiatives make up part of this complicated and ever-changing struggle for spoils. All these factors mean that Alberta is not operating in a vacuum when it comes to determining royalty rates.

Let's not forget that Albertans are also oil patch employees and citizens of Canada. And there's the rub. During a boom each of the players—citizens, politicians, oil companies, and investors—wants an increased share of the profits. As during the boom in the 1970s, everyone is lusting after the **windfall profits**. Will the Conservative government be able to devise a Canadian energy security strategy that will meet the needs of all Canadians?

The economic **booms** and **busts** regularly prove that Canada is a dominion of regions, each more interested in controlling and directing the development process than in co-operating with others. During the booms the level of stress is high, with particularly vicious attacks flowing back and forth between the petroleum-producing provinces and the consuming provinces. The federal government, for the most part, supports the largest number of voters—the consumers in Ontario and Quebec.

No wonder the West hates the National Energy Program. ∎

International Gas Prices—
A Comparison from Summer 2006

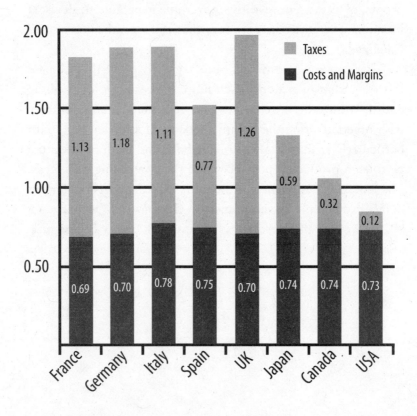

In Canada, taxes comprise about one-third of the retail price of gasoline.
This tax is lower than that in all other G7 countries, except the US.

Values Expressed in Canadian Dollars
Source: International Energy Agency

Outrageous
The Price at the Pump

Feeling ripped off at the gas station? Then be a wise consumer and buy your snacks in the grocery store. A self-serve gas station makes most of its profits from the pop, chocolate bars, potato chips, and other snacks it sells along with the gasoline. They operate on the same principle as convenience stores that sell milk at regulated prices and charge a lot for everything else, and movie theatres that sell competitively priced tickets to the show by charging you five dollars for popcorn that is worth a few pennies.

*The average Canadian drives
18,000 kilometres a year.*

We all drive cars—we're addicted to them, just like we are to many other conveniences like cable television and daily newspaper delivery right to the door. We don't have to feel bad about relying on vehicles any more than we do on other commodities that are important to us, but we can control our reliance on them.

In order to learn about the price at the pump you have to do some homework. Learn about it: what it costs, how to shop for it, how to use it wisely, and its alternatives. In other words, learn how to be a wise consumer.

The Facts

The average Canadian drives 18,000 kilometres (11,185 miles) per year. That's from Calgary to Vancouver nine times, from Toronto to Montreal seventeen times, from Ottawa to Quebec City nineteen times, or farther than from St. John's, Newfoundland, west across the Rock to the ferry, across all of Canada, and through the mountains to Whitehorse, Yukon, and back. Most of our driving is done close to home doing errands and spur-of-the-moment trips. They all add up.

It costs about 50 cents per kilometre, $9,000 a year, to drive a small car in 2007.

According to the Canadian Automobile Association, the fixed cost of running a Cavalier-sized four-cylinder North American car in 2005 was about $7,000 per year, including licence, registration, depreciation, and car loan. The variable expenses of tires, oil and gasoline, and maintenance added another 13 cents per kilometre—$2,340 for 18,000 kilometres. The result is 51 cents per kilometre, or about $9,340 per year, per vehicle. Most families drive two vehicles, and at least one of them is larger than a Cavalier. That's more than $25 a day per vehicle.

The price at the pump is just a small part of the cost of driving a car in Canada. It might cost you just $60 to fill up the tank in your car with 60 litres (15.8 gallons) of gasoline (when gas is $1 per litre), but the real cost to drive that tankful of gas is more like $306 when all car-related costs are included.

Cars are expensive. We rely on them extensively, and most of us are commuters. Most Canadians live in urban areas, and most city folk live in suburbs. Most of us also work far enough from our homes that we need to drive to work. Therefore, where we live and where we work both affect the cost of pulling up to the gas pump.

Canadians rely heavily on petroleum for heating our homes and fuelling our cars and trucks. When the price at the pump jumps, we feel trapped. It's easy to feel like we're addicted, and perhaps we are. The truth is that oil and gas are not like other commodities—you can't dig your own oil well, build a refinery, and make your own oil, gasoline, and natural gas, or create your own plastics. It can seem like we're trapped without choices, at least that's how it feels when we drive up to the gas pump.

A barrel contains 192 litres of oil, but only makes 85 litres of gasoline.

Fortunately, we do have lots of options. Gasoline is a commodity, and consumers can make informed choices, substitute alternatives, shop wisely, and even get their economic revenge on the faceless corporations that own the gas stations we visit every week.

A barrel of crude oil contains 192 litres (50.7 gallons). After refining that makes 85 litres (22.5 gallons) of gasoline, 40 litres (10.5 gallons) of fuel oil (for furnaces), 18 litres (4.8 gallons) of jet fuel, and 49 litres (12.9 gallons) of products like grease, asphalt, and source materials for plastics. Less than half of a barrel of oil becomes gasoline.

The cost of filling that barrel with crude oil includes exploration, drilling, field pipelining, and moving the oil to market through supertankers or large-diameter pipelines. Next comes the cost of refining and upgrading the gasoline to make it ready for the consumer. Provincial and federal taxes go on top of these costs. Finally, there are the costs of distributing the gasoline from the refinery to the gas station and running that gas station.

For a litre of gas selling for $1.04 in Toronto in June 2006, 49 cents went to pay for the crude oil. Ontario provincial tax was 15 cents, federal taxes were 17 cents, refining and wholesaling took 17 cents, and the retailer was left with 6 cents. Similar rates of taxes applied across the country, with some cities charging additional taxes to subsidize public transit systems to encourage car drivers to take the bus and train. There are other taxes, too. Property taxes include a hefty sum for roads, and provincial and federal income taxes also pay for highway construction and maintenance. It all adds up to a lot of money.

With lots of cash at stake, it seems like corruption could follow, but the Competition Bureau in Ottawa keeps a close eye on the oil industry and investigates any hints of wrongdoing. According to its commissioner, since 1990 the bureau has done five major studies into the gasoline industry and "found no evidence to suggest that periodic price increases resulted from a national conspiracy to limit competition in gasoline supply, or from abusive behavior by dominant firms in the market." In 1990, when Imperial Oil and Texaco merged in Canada, the Competition Bureau forced the newly expanded company to sell off a refinery, terminals, and hundreds of gas stations so that it did not become too large a force in Canada. In 1998 the Competition Bureau prevented the merger of Petro-Canada and Ultramar into one big company.

Although there are periodic and seasonal spikes in the price of gasoline, the bureau noted that prices drop after the peak season. It also found that high prices in the summer of

2004 reflected low gas supplies in North America and a short-age of crude oil worldwide. Overall, the commissioner noted that the international price of oil is a dominant part of the price of gasoline at the pump and that prices in North America are generally competitive.

Venezuelans pay the least for gasoline; Belgians pay forty-nine times as much.

Compared with producing countries where crude oil is cheap to pump out of the ground, gasoline prices in Canada are high, but when we compare the price at the pump with most nations, the ones that import oil, we have some of the lowest rates in the world.

Prices at the pump vary for many reasons. Disasters like Hurricane Katrina in 2005 can have a short-term effect. Katrina took more than 25 per cent of U.S. crude oil off the market immediately because of damage to coastal refineries and drilling platforms in the Gulf of Mexico. Crude oil prices rise due to storms, wars, or international crises anywhere in the world. OPEC sometimes gets blamed for the price of oil, though it only supplies 40 per cent of the world's oil supplies and usually boosts production during shortages. Besides, Canada is self-sufficient in oil production and supplies the United States with more oil than OPEC does. Prices change according to the season, and because we drive less in the fall and winter, the prices drop a bit after the busy summer season. If you are close to an oilfield your prices may be lower, and competition can affect the prices. Environmental programs also affect gas prices, and some regions try to minimize **smog** and other pollu-tants with stringent rules as to what goes into the gasoline in your area.

The Worldwide Price at the Pump in U.S. currency

Gas station (Courtesy Petro-Canada)

Least expensive per gallon/litre:

Caracas, Venezuela –	$0.12/3 cents
Lagos, Nigeria –	$0.38/10 cents
Cairo, Egypt –	$0.65/17 cents
Kuwait City, Kuwait –	$0.78/21 cents
Riyadh, Saudi Arabia –	$0.91/24 cents

Most expensive per gallon/litre:

Brussels, Belgium –	$5.91/1.56
Copenhagen, Denmark –	$5.93/1.56
Milan, Italy –	$5.96/1.57
Oslo, Norway –	$6.27/1.65
Amsterdam, Netherlands –	$6.48/1.71

Global Gas Prices (1 gallon = 3.8 litres)
(Prices as of March 2005. Source: CNN Money)

Global gas prices (Regular/Gallon)

Nation	City	Price (USD)
UK	Teeside	$5.64
Hong Kong	Hong Kong	$5.62
UK	Milford Haven	$5.56
UK	Reading	$5.56
UK	Norwich	$5.54
Germany	Frankfurt	$5.29
Denmark	Copenhagen	$5.08
Norway	Stavanger	$5.07
Norway	Oslo	$4.93
Italy	Rome	$4.86
Turkey	Istanbul	$4.85
Portugal	Lisbon	$4.80
Korea	Seoul	$4.71
Switzerland	Geneva	$4.56
Korea	Koje/Okpo	$4.53
Austria	Vienna	$4.50
Croatia	Zagreb	$4.32
Japan	Tokyo	$3.84
Australia	Sydney	$2.63
Cambodia	Phnom Penh	$2.57
Taiwan	Taipei	$2.47
Georgia	Tbilisi	$2.31
Laos	Vientiane	$1.66
Thailand	Bangkok	$1.60
China	Tianjin	$1.54
China	Shanghai	$1.48
Russia	Moscow	$1.45
Kazakhstan	Almatu	$1.36
Kazakhstan	Atyrau	$1.35
Tajikistan	Dushanbe	$1.32
Azerbaijan	Baku	$1.15
Venezuela	Caracas	$0.14

(As of February 2007, on money.cnn.com)

What You Don't See
Can Kill You—Smog

Smog is common in Canada's biggest cities, in the Windsor–Quebec City corridor in central Canada, and in the Lower Fraser Valley of British Columbia. But that doesn't mean all other areas are pollution-free.

Air pollution can be smoke from fires, wood-burning stoves, or industry. Tiny particles of pollution result from fossil fuels burned in our homes and vehicles.

Smog can hurt the eyes, nose, and mouth, irritate the lungs, and even cause heart problems. It affects the elderly more than young people, the sick more than the healthy.

We can avoid contributing to smog by taking public transit and joining car pools, riding bicycles, walking, and avoiding gas-powered lawnmowers and other machines. We can also make sure our vehicles are in good shape and fuel-efficient. An energy-efficient furnace helps too.

Smog is a byproduct of a lifestyle that relies heavily on burning fossil fuels, so reducing our dependence on them is good for the air and for our own health, too. ∎

You burn 673 calories riding a bike for an hour; an SUV burns 6 billion—okay, more like 93,000, but that's still a lot.

If you don't like supporting large oil companies, you do have options. For example, there is almost always an alternative to everything we purchase. You can buy Pepsi if you don't want to support Coke, or drink water if you don't like either of the big pop makers. If Microsoft is your favourite bad guy you can use another operating system, buy an Apple computer, or just use paper and pencil instead of a computer. You can buy soap from one of Procter & Gamble's competitors and boycott Wal-Mart if you don't approve of their way of running a business.

The Seven Sisters oil companies amalgamated into the Powerful Four: ExxonMobil, Chevron, Shell, and British Petroleum.

Though it seems like big oil has a monopoly, there are alternatives to buying gasoline from the four big oil companies: ExxonMobil, Chevron, Shell, and British Petroleum. You have lots of options. For example, you can buy your gasoline from an independent Canadian company or from a co-operative. You can convert your car to propane or natural gas, or even buy a hybrid vehicle that sips energy even more slowly. Also, smart consumers shop around, thus helping to diversify our suppliers. We operate within a budget and make sure that our spending habits reflect our priorities in life and support our values. We apply these principles when purchasing a home, a holiday, a television, or a university education. The same applies to gasoline.

The Profit Game

Oil company profits are always a cause for suspicion. However, according to a report by Bloomsberg News in 2005, banks made a 33 per cent profit, Microsoft recorded 32 per cent, Coca-Cola earned 21 per cent, Procter & Gamble managed 14 per cent, General Electric and ExxonMobil were at 11 per cent, ConocoPhillips was at 8 per cent, IBM and Chevron were at 7 per cent, and Wal-Mart posted 4 per cent. ■

Our Options

As consumers of transportation we have lots of options. We can think of the cost of commuting when we make major life decisions, such as buying a home and taking a job. If your home and place of work are far apart, commuting will be a significant cost in both time and car expenses, so you might want to move closer to work or to a high-speed transit system that gets you to your job quickly.

Most families need one car, but do you really need two? And a truck or a motorhome? Some options include using the bus or metro, a taxi, ridesharing, renting a second car a few times each year, or getting really creative and walking, running, or biking part of the time. Maybe your lifestyle requires a car and a larger vehicle like a truck. Fine, but remember that you made decisions about where to live and work, how to get around, and where to go on holidays that affect the number of vehicles you have to support. Hybrid cars that use electric engines and small gasoline pickups are an option, as are vehicles with small diesel engines. Carpooling works too, or joining a car co-op and sharing the ownership of a vehicle with others, an option available in most large cities in Canada. Hydrogen vehicles and ones that rely totally on electricity are still just prototypes; these may never end up in your driveway—or they may!

Do you really want to spend one-third of your income on cars?

Budgeting is a nuisance, but if you apply the real cost of 51 cents per kilometre to your driving and see how much it really costs, you might feel motivated to make some changes. If it costs you $20,000 to keep two vehicles on the road and your family income is only $60,000, then spending one-third of your income on cars is probably too much, but if you make $100,000 then spending one-fifth might make sense.

Your stage in life probably affects your transportation needs. Many singles or young marrieds can get by without a car in a city, but once the babies come along, a car is a must. With busy teenagers in the family two cars might be necessary, so one parent can be a taxi driver for a few years, ferrying kids around to different schools as well as sports and social events. Empty-nesters sometimes cut back to a single vehicle or move into an apartment and use public transit.

You can also use your purchasing power to maximize your gas costs. Find out where gas is cheap in your city. Join a co-op and get a discount—most parts of Canada have consumer co-ops where members get a significant discount for purchasing gasoline through the company. Get a fleet card, use self-service, or buy on Wednesdays, Saturdays, and Sundays—some gas stations offer discounts on set days of the week. Buy your gasoline with a credit card that rewards you with cash, air miles, or some other incentives.

You are not alone in this quest. A search on the Internet in 2006 found lots of other ideas for how to save at the gas station.

Using less is a good place to start. Only put $20 of gas in the tank so that it is always nearly empty—a motivation to walk or use the vehicle carefully—and keep the car in good repair. Slow down to 90 kilometres (56 miles) per hour, buy a hybrid or a diesel, and do all shopping for the week during one trip to a large supermarket. You could also consider riding a bicycle, buying a small scooter, trading the SUV for a small, efficient car, or working from home. Stop buying expensive bottled water at the gas station convenience store, and stop using drive-in windows at restaurants and banks. Vote wisely and write to your politicians at each level of government to tell them what you think about these issues. Don't feel bad about your habits, just choose better ones and look for ways to be more responsible. Even oil companies are looking for efficiencies in their operations, for ways to use less oil and gas.

If you still feel ripped off even after reducing the amount you drive, cutting back to one efficient car, and shopping wisely, you can get economic revenge on the oil companies in another way. Buy their stock. By investing in an oil company you become an owner, and you have the right to request investor information. That way, you'll be getting part of the profits that you paid at the pump.

Finally, don't buy your snacks at the gas station. ■

Pumped.
(Courtesy David Finch)

Key Dates

1875 –   Geological Survey of Canada considers
                  commercial viability of oil sands

1915 –   Oil sands shipped to Ottawa and used to
                  pave Wellington Street

1925 –   Karl Clark first separated bitumen from oil
                  sands in a laboratory

1967 –   Great Canadian Oil Sands (Suncor) plant
                  began producing oil from oil sands

1970s to 1990s – Production efficiencies allow price per barrel
                  to drop from $35 to $13

1999 –   Oil sands and heavy oil wells produced as
                  much oil as conventional wells

2004 –   Daily production of oil sands reached one
                  million barrels

Bitumen
The Oil Sands

W e can't call them '**tar sands**' anymore," said Karl Clark and Sid Blair, the pioneers in the complex job of removing oil from sand. "Let's call them 'oil sands' instead!" And so, in 1951, the Research Council of Alberta cleaned them up—no more reference to sticky, gooey tar.

Call it "tar" sands or "oil" sands, it's all bitumen or heavy oil, a black, tarry source of hydrocarbons as thick as molasses on a cold day. Scientists believe the petroleum in the oil sands probably came from the same unknown source as all the rest of the oil in the province. It then moved east, pushed by the forces that created the Rocky Mountains. In the northern half of Alberta, sand in ancient channels of the Peace and Athabasca rivers trapped the oil. The same thing happened to the sands in the Cold Lake area. For thousands of years the oil remained an untapped resource.

Daniel Diver boiled tar sands in his shack near Fort McMurray and produced natural gas, oil, and sand in March 1920. (Glenbow Archives, NA-1142-6)

We've known about bitumen for centuries. The Egyptians used it to embalm their dead, and the Bible says Noah used it to pitch his ark. Even the Tower of Babel relied on its adhesive powers to serve as mortar for its bricks. Early Europeans used it to make tools, and it was used throughout the world for waterproofing buildings and boats.

Oil sands lie under about 23 per cent of Alberta.

Natives in northern Canada pointed out bitumen to Henry Kelsey of the Hudson's Bay Company in 1719, and wild man Peter Pond saw oil sands deposits on the Athabasca River near

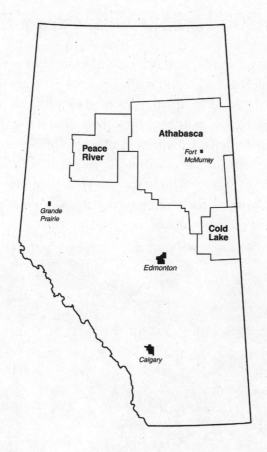

Alberta's oil sands. (Map by Toby Foord)

today's Fort McMurray in 1778. Alexander Mackenzie noted oil seeps along the Mackenzie River in 1789. He wrote about pools of the stuff "into which a pole of twenty feet long may be inserted without the least resistance. The bitumen is in a fluid state, and when mixed with gum, or the resinous substance collected from spruce fir, serves to gum the canoes." But without any use for it, traders and explorers paid little attention to the mysterious sands that seeped oil into the river.

From 300 to 55,000 Barrels a Day

Each glob of oil sand contains a core of sand, then water, and finally bitumen or oil around it all. The trick is to get the constituents apart. Modern attempts to produce oil from bitumen began in the 1920s, and in 1930 Robert Fitzsimmons of the International Bitumen Company built a small plant that used hot water to separate the oil from the tar sands at Bitumount, on the banks of the Athabasca River near today's Fort McMurray. Working hard, seven men managed to produce just 300 barrels of oil that summer.

Dr. Karl Clark testing the process in the Tar Sands Department at the University of Alberta in Edmonton, 1929. (Glenbow Archives, ND-3-4596a)

Others tried to develop the resource in the 1930s, including Max Bell and B. O. Jones, who formed Abasand Oils Ltd. They eventually built a tar sands plant at Fort McMurray and in 1945 announced that commercial production problems had been "licked."

In 1951 Alberta mines minister N. E. Tanner, announced new regulations to help companies exploit Alberta's "fabulous and so far uneconomic" Athabasca tar sands. As a result, ten Canadian companies leased land from the province in 1952 and promised to begin **core drilling** programs that spring. Sun Oil Co. of Pennsylvania also got in on the deal and took out a lease on 100,000 acres (40,470 hectares) of land.

It takes 2 tonnes of oil sand to make one barrel of oil.

Through the 1950s many other operators announced plans to strip the oil from the tar. The most unusual project was in 1958 when Richfield Oil Company revealed "**Project Cauldron**," an ambitious plan to detonate a nine-kiloton nuclear bomb to melt the oil deep underground. As ridiculous as this proposal sounds, it received support not only from the American oil company but also from almost every level of the Alberta government and its civil servants. Only Dr. D. Dick of the Department of Public Health seemed to hesitate: "It would seem that the setting off of a nuclear device beneath the Oil Sands in Northern Alberta should pose little or no hazard to health or safety ... but what constitutes a safe level of radiation is not yet known." Today's plan is to use a nuclear reactor to make the heat needed to process the tar sand, instead of natural gas.

Though many scientists and a large portion of the public were infatuated with nuclear weapons and the potential for putting them to peaceful use, support did not come from all regulatory agencies. Ottawa's conditional approval for the project died when the prime minister appointed Howard Green as the new secretary of state for external affairs in 1959. Green supported nuclear disarmament and hoped Canada would play an important role in freeing the world of the destructive weapons. As a result, Alberta pulled back its support for the nuclear project, though Richfield continued promoting it until 1963.

While separating the oil from the sand was technically feasible, it took until 1967 for commercial production to begin at Fort McMurray. In 1960 Sun Oil announced plans for a $110-

Historic Crude Oil Prices

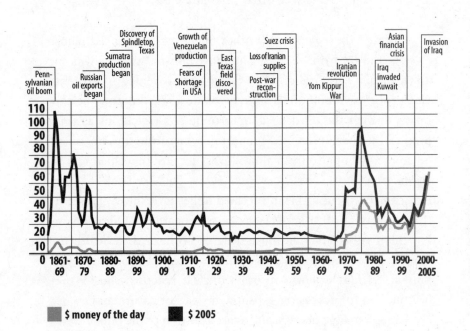

$ money of the day $ 2005

million Great Canadian Oil Sands (GCOS) project, but later that year the **Alberta Oil and Gas Conservation Board** (Energy Resources Conservation Board—the ERCB) rejected the project, questioning its economic viability. Two years later the Conservation Board gave GCOS the green light.

Low oil prices made the oil sands companies improve efficiency.

The early 1960s was not a good time to be adding more oil production in Alberta. In 1963, due to a glut of oil in the North American market, the Conservation Board refused to approve permits for Shell Canada and Cities Service to develop oil sands projects. Construction began on the GCOS oil sands project at Fort McMurray in 1964, and the estimated cost of the facility grew past $190 million.

In June of 1967, as the GCOS project was nearing completion, war broke out in the Middle East. A dozen scuttled ships blocked the Suez Canal, always a bottleneck in the shipping of oil from the Middle East to North America. Tankers then had to go around Africa, delaying delivery of oil and threatening security of supply. Oil prices rose overnight. In August of that year Arab leaders considered a worldwide oil embargo to illustrate their power and ability to control the world oil market.

The GCOS plant officially opened shortly after the Suez Crisis, on September 30, 1967. J. Howard Pew, chairman of Sun Oil, stated, "No nation can long be secure in this atomic age unless it be amply supplied with petroleum ... It is the considered opinion of the Sun group that if the North American continent is to produce the oil to meet its requirements in the years ahead, oil from the Athabasca area must of necessity play an important role."

Premier Ernest Manning of Alberta said, "This is a red letter day, not only for Canada but for all North America. No other event in Canada's centennial year is more important or significant."

Later renamed Suncor, this pioneering company is still a lead player in the development of oil sands in Alberta. Another company, Syncrude, a consortium of the federal government and several oil companies, dates from the 1960s. Formed in 1964, it took over the Cities Service Athabasca Inc.'s tar sands research program and its oil patch partners. Governments eventually became involved, investing in the project. Construction began on a second major oil sands plant in 1973. Oil began flowing out of the Syncrude plant at a rate of 55,000 barrels per day in 1978, and in 1998 it celebrated the production of its billionth barrel of oil.

In the early 1970s the ongoing tensions between Alberta and Ottawa over control of the oil industry came to a crisis. Oil-consuming provinces, most notably Ontario, argued against any higher oil prices. Premier Lougheed of Alberta argued his province's case in the *Edmonton Journal* on June 24, 1974: "We should not forget, nor should the rest of Canada, that no other province is expected to put its resources on sale at the special rate for other provinces." Prime Minister Trudeau grudgingly agreed with the Alberta position, admitting that declining reserves and production made it necessary to allow the producers a larger share of oil revenues to enable them to search for more petroleum. "We cannot expect those who search for oil—whether they be Canadians or others—to look for it and develop it in Canada if our prices are far below those in other countries ... So my colleagues in the government and I have come reluctantly to believe that the price of oil in Canada must go up—up towards the world price. It need not go all the way up."

But by the end of 1974 one part of the industry had indeed

convinced both Alberta and Ottawa to allow it to charge the world price for oil and to exempt it from many of the new taxes and royalties that tapped into the petroleum wealth. Syncrude, Canada's second largest oil sands project, threatened to collapse during 1975 after partner company Atlantic Richfield withdrew its support for the project. Since Ottawa, Ontario, and Alberta had all counted on this new megaproject to yield jobs and a secure petroleum supply, the remaining partners in the project—Imperial Oil, Cities Services, and Gulf Oil—used Atlantic Richfield's withdrawal from the project to force both levels of government into granting unprecedented concessions. In the end, Alberta, Ontario, and Ottawa all became financial partners in the project. Alberta also paid infrastructure costs including a $300-million utility plant as

What They Do...
Oil Sands Workers: Turning Mud to Gold

In the early days tar sands workers had to shovel the gooey sand onto barges and haul it to the processing plant upriver. In Edmonton and Ottawa they even paved streets with the raw bitumen.

Today's oil sands workers convert tarry muck to gasoline using more sophisticated machinery. But it takes more than truck drivers and heavy equipment operators to extract petroleum from the oil sands. It also takes estimators, accountants, boilermakers, quality control specialists, millwrights, mechanical, geological, geotechnical, and mine-planning engineers, operators, pipefitters, planners, analysts, buyers and procurement professionals, electricians, carpenters, administration assistants, trainers, managers, drillers, welders, estimators, human resources advisers, environmental specialists, and many others.

Most of these careers require specialized university or technical school education. ∎

well as a $100-million pipeline to Edmonton from Fort McMurray. It built community schools, bridges, highways, and other services. Syncrude received the world price for its oil, and its private corporate partners received generous **write-offs**, not only on expenses directly related to the oil sands plants but also on exploration and development projects in other parts of their operations. In the end the corporations paid only about 25 per cent of the capital costs of the $2-billion project, but gained title to 70 per cent of the project, while the federal and provincial governments paid 75 per cent of the costs and received only 30 per cent of the ownership.

The Syncrude story proved that when pushed hard enough, both the provincial and federal governments would grant the oil industry full world price for oil, as well as generous tax exemptions and other provisions. It also shows that oil sands development is expensive, prone to cost overruns, and creates large demands on social and other services in the communities near the developments.

Getting the Sand Out of the Oil Sands

Thirty years later the oil sands are even more important to Alberta, Canada, and the United States. With supplies of conventional oil and gas decreasing quickly and North American consumers using more petroleum every year, the buried treasure in the bitumen may be the promise of the future. About two trillion barrels of oil are trapped in the sand, enough to supply everyone in North America for decades, maybe even for hundreds of years. Only Saudi Arabia has more oil reserves than Canada, and those supplies are half a world away from North American oil consumers.

Only 12 per cent of the oil can be extracted from the oil sands with current technology.

As of 2007 more than 60 per cent of Alberta's oil production, and about 40 per cent of Canada's oil, came from the oil sands, totalling more than a million barrels every day, or about 10 per cent of the total North American oil supply. The amount is growing every year. Current plans call for four or five million barrels a day by 2020, and maybe more. Dozens of corporate investors have almost one hundred projects in planning. The technology, demand for oil, and high price all point to incredible growth.

But to get one barrel of oil, 2 tonnes of tarry sand have to be dug out of the ground by big shovels and loaded into huge 380-tonne trucks that take it to a plant to be processed. Some bitumen goes to the plant through a pipeline, after being broken

Tar sands operations near Fort McMurray, Aberta, 1936.
(Glenbow Archives, NA-3394-57)

up and mixed with hot water. Only 90 per cent of the oil comes out of the sand, so the rest is returned to the ground. When the oily sand gets to the plant, hot water helps release the oil from the sand as the whole mixture gets shaken and stirred. Bitumen rises to the top and goes off for further processing, and more oil gets squeezed out of the rest of the slurry. Even then, it still has to go through an **upgrader**, which makes it lighter using **diluents** or lighter oil, before it can go into a refinery to create "synthetic" crude oil.

Bitumen is not like other petroleum products, so it has to be upgraded, a process that removes some of the carbon and adds more hydrogen molecules. These chemical changes are why it's called **synthetic oil**. Processed sand goes back into the pits while the water, contaminated with sand, clay, and bitumen, goes to **settling ponds**—large pools dug into the ground. These **tailings ponds** covered an area of 50 square kilometres (19.3 square miles) in 2005. Some water from these ponds goes back into the refining process, but the suspended particles take many years to fall out of the water. Each barrel of oil produced releases 80 kilograms (176.4 pounds) of **greenhouse gases**, a product known to contribute to climate change during the process of extracting it from the oil sands. When the oil becomes gasoline and we burn it in our vehicles, it emits up to five times more greenhouse gases than in the production process.

Only a small portion of the oil sands is close enough to the surface for us to reach with heavy machinery. Where it is buried by more than 75 metres (246 feet) of earth—the length of an NHL hockey arena—it has to be coaxed out with other techniques. Currently only a third of the production from the oil sands comes from underground operations, through processes that began commercial operation in 1985. In some situations the oil sand will even flow to the surface on its own. This is called the **cold production** method, but it produces less than 10 per cent of the oil in the field. Most bitumen is so thick that it requires persuasion. Heating the bitumen underground can cause some of it to liquefy and move to a well.

Using steam in a variety of ways also recovers the oil *in situ,* and it could help produce up to 80 per cent of the total oil in the reservoir. The huff and puff steam method, or **cyclic steam stimulation** process, forces 300°C steam into a layer of bitumen for several weeks, and then pumps suck out up to 25 per cent of the oil from the same wells that injected the steam. **Steam assisted gravity drainage (SAGD)** forces steam horizontally into the bitumen through one well and then sucks out up to 60 per cent of the oil through another well 5 metres (16.4 feet) lower down in the formation.

Other methods used to extract oil *in situ* include the **vapour recovery extraction (VAPEX)** method, which uses carbon dioxide or solvents, sometimes with steam, to coax the oil to the well. In addition, **Toe to Heel Air Injection**, or **THAI**, is being tested. This method uses **fireflooding**, which injects oxygen into the formation and then ignites the bitumen with the resulting heat melting out the oil, but it also sometimes sets fire to the wells that are pumping out the oil.

Most of these systems require water and natural gas—lots of it. It takes between two and six barrels of water to separate one barrel of oil from the bitumen, depending on the processing. Some water goes back into the plant, but most of it goes out into tailings ponds. **Caustic soda,** used during processing to help separate the oil from the sand, can cause clay to stay suspended in the water indefinitely, so some systems avoid using soda, and others use **gypsum** to help settle out the solids. Using non-drinkable or **brackish water** is another option.

The *in situ* operations generally use more water than the mining process, and 20 per cent of the water stays underground. In 2005 oil sands companies were licensed to take a maximum of 350 million cubic metres (over 12 billion cubic feet) per day out of the Athabasca River, or as much water as a city of two million people uses each day. The oil companies are still trying to find ways to reduce the amount of water used per barrel of oil sand, but the plans to greatly increase production from this source mean that even more water will be required in the future. Neither industry nor government has a solution to this problem at this time.

Boreal forest along the Coppermine River, Nunavut.
(Courtesy James Raffan)

The Oil Sands Water Issue

If Alberta oil sands production is to continue to expand and eventually produce up to five million barrels of oil per day—five times the 2005 rate—then the water issue deserves more attention and careful planning. The National Energy Board takes the issue seriously and states: "Oil sands mining operations use large amounts of water, and a variety of regional multi-stakeholder groups agree that the Athabasca River does not have sufficient flows to support the needs of all planned oil sands mining operations."

Though less than 1 per cent of the total *annual* flow of the Athabasca River is used by the oil sands companies, it still amounts to 65 per cent of the water taken out of the river. The rate at which industry has been increasing its use of water has been growing so quickly that between 2001 and 2004 it tripled the Alberta government estimates. The oil sands will soon use more water than Toronto.

Although water is reused when possible, most of it does not return to the river and ends up in tailings ponds. Water used for *in situ* production underground does not return to the environment because it is trapped underground. Some operations use salt water in order to reduce the consumption of fresh water; others use solvents instead of water. Also, conventional oil production is declining, which means that less water is used to produce that resource.

Critics point out that limits to water use are voluntary, not compulsory, and that industry does not pay for the water it uses; charging for the supply would help encourage conservation. Waste water treatment and disposal are also causes for concern. New technologies are in the works but may not become available before 2030.

The National Energy Board also notes that winter water draws "could impact the ecological sustainability of the river." Government agencies will need to continue to closely monitor water use, supplies, and the effects of water use on the environment. Water use may have to be restricted to less than 10 per cent of the current flow of the river in order to prevent negative impact on the river. New projects may be required to build thirty-day water storage facilities on site to deal with times of low water supply from the river. The cumulative effects of development on water resources, both river and groundwater, are still unknown. ■

Natural gas is an important part of the oil sands recovery process too. During the 1990s and the early part of the twenty-first century Alberta expanded its use of natural gas by more than 60 per cent. It takes between 750 and 1,500 cubic feet of natural gas (enough to heat a house for up to a week in the winter) to heat the water that helps separate the oil from the sand, and that much again to upgrade the bitumen into oil. Each day the oil sands operations use enough natural gas to heat 3.2 million homes for that day.

Disturbing the Boreal Forest

Environmental concerns surrounding oil sands operations are many. Mining always involves disruption of the landscape, sometimes for generations. The development process also affects the forests, plants, and animals. Oil sands companies have been working with the government to remove the **overburden**, or top-soil and plants, and store it for use in the **restoration** phase, when they promise to renew it as best as possible. Concerns include the size of the mined property, the salts and other minerals that go back into the pit with the tailings, and the cost of restoring the land if the price of oil drops or if companies go out of business. Full restoration is impossible because it would involve recreating the whole ecosystem, including the peat, the natural plants, and the animals that lived on the land for thousands of years.

One barrel of oil makes enough gasoline to move a large SUV 500 kilometres—about the distance from Fort McMurray to Edmonton.

Some of the mined property at Fort McMurray has been filled back in, replanted with grass and trees, and some buffalo are grazing there, but none of the land undergoing restoration work has been certified as "reclaimed" by the Alberta government. According to the **Energy Resources Conservation Board (ERCB)**, "**Reclamation** of mining sites is done to a standard to at least the equivalent of their previous **biological productivity**." In 2006 the Alberta Energy website clarified the issue and added that "reclamation will not replace exactly what existed prior to mining, but will create a self-sustaining boreal ecosystem with a natural look that fits with the adjacent landscape." Perhaps the government's reluctance to classify any land as reclaimed is because it will be impossible to successfully return the land to its pre-mined condition as boreal forest. The **National Energy Board (NEB)** noted in 2006, "It is uncertain if land reclamation methods currently employed will be successful." Critics wonder if biological productivity can be restored and are concerned about the cumulative effects of the fast-paced development. For their part, the companies are doing massive and continuous reclamation of thousands of hectares each year, and in some cases have reduced the time the land is in the mining stage from thirty years to about ten years. Though each project files an **Environmental Impact Assessment** with the government, the full effects of tailings ponds, greenhouse gases, and other emissions are not fully understood.

The environmental challenges presented by the oil sands development process are huge, and it is in the best interest of all Canadians to deal with them successfully. Keep your eye on the media to see what industry and the government are doing to make sure that the natural environment is being restored.

To Be Resolved

The wealth generated by the oil sands development process makes jobs and puts money into government treasuries, but royalties from the oil sands are different from other oil and gas operations. Because of the high cost of developing the resource—oil sands costs about $15 per barrel to produce compared with $2 for Middle East oil—the Alberta government gives oil sands companies a break on royalties for many years. Until profits pay for the construction phase, the government will only get 1 per cent of gross sales royalty, compared to about 25 per cent for conventional oil and gas. Once the construction phase is paid for, the rate will change to 1 per cent of the gross revenue or 25 per cent of the net revenue. The current agreement dates to 1996, when the price of oil was just $30 per barrel. Though oil sands royalties have earned the provincial government more than $4 billion since 1990, critics argue that the 1 per cent rate was set when the world oil price was low and that it should be renegotiated to reflect the higher price and strong demand for petroleum from this public resource.

By 2015, three of every four barrels of Canadian oil will come from the oil sands.

Quality of life is another public resource that can suffer during a booming development period. Though jobs pay well in the oil sands, the cost of living is high in nearby communities. For example, housing costs went crazy in 2005, with the average price for a home in Fort McMurray topping $450,000. A comparable home in Edmonton cost $220,000. Rents in Fort McMurray were the second highest in Canada, with a one-bedroom apartment going for almost $1,100 per month. All public services in the city were burdened too, straining to cope with the demands of thousands of new immigrants each year.

Transportation systems are also suffering. Highway 63 between Fort McMurray and Edmonton is very busy. Traffic flow increased 140 per cent between 1996 and 2004, and the booming construction in the area promises even more demands on this one highway. Twinning the road is a high priority, but the cost will fall to the taxpayers.

A railway line links the oil sands area to Edmonton, but it does not carry passengers. Two small companies operate sections

Glenn Fox, Bob Sharp, Jack Sinclair, and Con Hage in geological survey fly camp, 1941.
(Glenbow Archives, NA-3834-3)

of the line, both moving freight over their part of the 507 kilometres (315 miles) of track between Edmonton and Fort McMurray. A train car carries as much as three transport trucks. The old track cannot support great speeds, so the trains chug along between 20 and 50 kilometres (12.4 to 31 miles) per hour. While some dreamers imagine building an expensive railway to replace the current system, loads of equipment and supplies move every day between the capital of Alberta and the oil sands.

Barges may offer another way to get heavy loads to the oil sands. In 2006, after a twenty-seven-year lapse, tugs once again started moving south along the rivers from Great Slave Lake to the Fort McMurray area. Common routes for moving freight *to* the North during previous oil booms and World War II, the rivers and lakes may soon carry freight *from* the North. Tugs and barges can move large equipment built overseas up the Mackenzie River, across Great Slave Lake, and then up the Slave and Athabasca Rivers to Fort McMurray and other oil sands communities. One portage is required, a 26-kilometre (16-mile) detour around the rapids between Fort Smith and Fort Fitzgerald on the Slave River, but in July 2006 a 160-tonne tugboat and a large barge proved it could be done. The route could be in regular operation by 2009.

Plans call for the oil sands to generate up to five million barrels of oil per day.

After four decades in production, the Alberta oil sands have become the single largest source of oil in Canada. Only the Middle East has more oil. There is enough oil in these deposits to build a four-lane highway to the moon—400,000 kilometres (248,555 miles). Though plentiful, oil sands oil is also hard to separate from the sand and more expensive to produce than oil from most conventional sources. According to the National Energy Board, current production methods can only recover 12 per cent of the petroleum in the oil sands.

Still, the oil sands have the potential to provide a long-term supply of oil for North American needs. Successful development of this massive resource requires ongoing investment from private companies and many levels of government. Challenges will continue to demand creative solutions to the technological, economic, political, and environmental aspects of this enormous enterprise.

Back in 1875, John Macoun of the Geological Survey of Canada looked into the future as he wrote in his diary on the shores of the Athabasca River: "Long after the noises [of camp] ceased I lay and thought of the not far-distant future when other sounds than those would wake up the silent forest; when the white man would be busy, with his ready instrument stream, raising the untold wealth which lies buried beneath the surface, and converting the present desolation into a bustling mart of trade."

That sound is in our ears. ∎

Key Dates

1719 – Cree Wa-Pa-Sun showed tar sands to Henry Kelsey of the Hudson's Bay Company

1857 – First oil produced at Oil Springs, Ontario

1859 – Natural gas flared as waste in New Brunswick

1883 – Water-well drillers discover gas at Langevin, Alberta

1902 – Oil discovered in what is now Waterton Lakes National Park, Alberta

1914 – Wet gas discovery at Turner Valley, Alberta

1920 – Norman Wells oilfield discovered in Northwest Territories

1943–45 – First offshore well drilled on artificial, unnamed island off Prince Edward Island

1966 – First well drilled in Beaufort Sea

1967 – Sable Island gas discovered by Shell

1967 – First drilling off Canada's West Coast

1971 – Taglu gas discovery in Mackenzie Delta, Northwest Territories

1973 – First oil discovered off East Coast at Cohasset in the Atlantic Ocean

1985 – Panarctic shipped first oil from Bent Horn, Arctic Islands

1997 – First oil production from Hibernia, Grand Banks, Newfoundland

2000 – First gas production from Sable Island

2002 – Coal bed methane plans announced for Canada

2002 – Gas hydrates experimental project started in the Mackenzie Delta

Frontiers
Oilfields on the Edge

"The sky is falling! The sky is falling!" Chicken Little gets busy every few years, running around yelling that we are running out of oil and gas, but we always have frontiers to exploit. For nearly four hundred years the almost unlimited resources of the Canadian landscape have provided new sources of raw materials. Timber, fur, coal, grazing land, and the agricultural land of the prairies—you name it, we have been lucky. We have been really lucky with oil and gas.

We are not running out of petroleum, not by a long shot. When scarcity seemed to be threatening Canadian oil and gas resources in the early 1970s, industry leaders went to Ottawa to share their knowledge.

Running Out of Cigars

Cal Evans and George Grant, past presidents of the Canadian Society of Petroleum Geologists, updated the House of Commons Standing Committee on National Resources and Public Works in 1973. In spite of the worldwide petroleum crisis caused by the Middle East War and the Arab oil embargo, Canada had lots of reserves. Oilmen who had served as president of the Canadian Society of Petroleum Geologists knew that Canada was not running out of petroleum reserves.

"So we weren't running out of cigars," Evans recalls. "We were just running out of 5-cent cigars." Many of the oilfields that we are exploiting today were already identified in the 1970s. It's just that they were far too expensive to develop when the price of oil was barely $10 per barrel.

Oil prices, predicted to rise to $100 per barrel during the 1980s, fell to a low of less than $19 in 1988.

The price of oil skyrocketed to nearly $45 per barrel during the early 1980s, and pundits predicted it would inevitably rise to $100. But it didn't, and the worldwide economic downturn in the mid-1980s caused the price to fall to less than $20 per barrel.

Each time natural resources lose their market value we have to find an alternative source, create a way to extract them more cheaply, or reinvent the way we use seemingly useless resources. For example, today's innovators are developing techniques for liberating natural gas that is trapped on the ocean floor far out at sea. They are also producing gas from the deep coal formations we no longer exploit. Most coal mining done currently in North America comes from open pit mines.

Historical Frontiers: The West and the North

Hundreds of years ago explorers from England and continental Europe came to North America looking for resources. Fish and timber were the earliest success stories, though the would-be heroes were looking for gold and silver. As they and their Native trading partners exploited the inland areas, they developed the fur trade too. Later came settlers, farmers, and others, all living off the land. Along the way they noticed other natural resources, but few of these resources made much money after being shipped across the Atlantic to distant markets.

Just as North America was a wealthy frontier for the Old World to exploit, much of the land west of Ontario was considered by Upper and Lower Canada—today's Ontario and Quebec—to be the frontier. When the developed lands ran out, people just moved west.

Frontiers can be geographical, technological, economic, and political.

The explorers of the eighteenth and nineteenth centuries always went far ahead of the settlers, looking for resources their companies or governments might develop. Henry Kelsey, Peter Pond, Alexander Mackenzie, and Henry Moberly noticed the tar sands when on voyages of discovery through the region.

In 1875 the Geological Survey of Canada (GSC) sent botanist John Macoun out west to investigate the northern rivers, and he once again noted the tar sands. Dr. Robert Bell of the GSC further studied them in 1882, and the next year G. C. Hoffman of the same department successfully separated oil from the bituminous sands with hot water.

While drilling for water for its steam engines, the Canadian Pacific Railway hit natural gas in 1883 at Langevin, near today's Medicine Hat, Alberta. British author Rudyard Kipling, best known for the children's story *The Jungle Book*, later described the city as having "all hell for a basement," and the plentiful supply of natural gas has given the region a natural economic advantage for generations. The gas was so plentiful and cheap (free!) that its streetlights began burning natural gas day and night in 1900.

John George "Kootenai" Brown, circa 1910.
(Glenbow Archives, NA-678-1)

The next frontier discovery was also in the West. Based on information passed on by Natives to Kootenay Brown, drillers punched an oil well into the ground in today's Waterton Lakes National Park, Alberta. The 1902 "discovery well" produced 300 barrels of oil per day, and Oil City sprang up as a result, but commercial production at this site never made the Waterton oilfield financially viable.

Transportation is often a key challenge when exploiting frontiers.

In 1914 wet gas greeted drillers at the Dingman No. 1 well in Turner Valley, Alberta, just southwest of Calgary. The overnight boom resulted in more than five hundred companies being formed, most of which were not worth the paper their stock certificates were printed on. As the story goes, disappointed investors used the worthless oil company stocks to paper the inside of their outhouses. But Turner Valley oil discoveries eventually caused two additional booms, from 1924 to 1930, and again from 1936 to 1945. Based on the oil and gas in these oilfields, multinational companies and Canadian firms continued the search for oil in Alberta and the West throughout the early decades of the twentieth century.

Another significant discovery occurred in 1920 far up the Mackenzie River in the Northwest Territories. Based once again on Native sightings and surface seepages, drillers struck crude oil at Norman Wells, roughly halfway between the Alberta border and the Arctic Ocean, but it was too far from markets. Nearby consumers used it at mines, and it served the war effort briefly during World War II, but it was not until 1985 that oil from the Far North began making its way to southern Canada through a pipeline.

The oil-bearing reefs at Leduc are up to 400 million years old.

The next significant discovery in the West was in 1947 at Leduc, near Edmonton. Though not remote, the discovery of this frontier was on the edge of geological knowledge, and it exploited the oil found in Devonian-aged reefs buried deep below the Alberta prairies hundreds of millions of years ago. By the end of 1947 several dozen wells were producing more than 2,500 barrels of oil per day. This new geological find allowed geologists throughout the Canadian West to review their previous research and find many new oilfields and drill thousands of successful wells. Though many other discoveries followed, the discovery of oil at Leduc in 1947 signalled the start of the richest period in Alberta oil history. It prompted companies from around the world to search for oil and gas in western Canada and provided enough oil (annual production mushroomed from less than 7 million barrels in 1946 to 144 million barrels in 1956) to allow for the construction of the first pipelines to move Alberta petroleum to markets in central Canada and the West Coast of Canada and the United States. Leduc made Canadian oil a continental resource.

Historical Frontiers: Ontario, the East, and Offshore

Oil seeping from the ground also attracted interest in the frontier region of western Ontario in the 1800s, and the first oil wells in North America began exploiting the resources at Oil Springs, Ontario, in 1857. Sour gas also greeted drillers in this area, leading to the development of a natural gas industry and

The Devonian

Until the late 1940s, all Alberta's oil came from relatively shallow oil wells in the Mississippian formations that were formed 360 to 325 million years ago. So it was a big shock in 1947 when drillers hit oil in a Devonian reef in much older rock formations that date back to the "Age of Fishes." The Devonian Period of the Paleozoic Era was from 416 to 359 million years ago. These rocks trapped immense quantities of oil as it migrated through the area. The town of Devon in Alberta is named after this discovery. ■

Leduc-era sample catcher Bob Schwarz shows places in Devonian rock sample where oil was trapped. (Courtesy David Finch)

the first exports of gas to the United States by pipeline in 1891.

In New Brunswick, Dr. H. C. Tweedle found both oil and gas near Moncton in 1859, but water seepage prevented production from these wells. For a time they just flared the natural gas, but in 1912 a pipeline linked the Dover gas field to Moncton, and residents of that city enjoyed natural gas from this early field for more than half a century.

Most oilfields give up less than
15 per cent of their oil.

On the East Coast, a 4,267-metre- (14,000-foot-) deep well produced some oil in 1942 from an artificial island 10 kilometres (6.2 miles) off Prince Edward Island. Farther offshore, Sable Island gas burst into the public imagination in 1967, but it took until 2000 for commercial production to begin from that source. Offshore drilling began in the Arctic in 1966 with a first well in the Beaufort Sea. Drilling began off Canada's West Coast in 1967, but for environmental reasons (reviewed in the chapter OOPS!—Mistakes and Lessons, pp. 137–61), the federal government placed a moratorium on petroleum production in the Pacific. The Taglu gas discovery in the Mackenzie Delta of the Northwest Territories in 1971 added more potential, as did the 1973 oil find at Cohasset, just south of Sable Island in the Atlantic Ocean. In 1985 Panarctic exported its first oil shipment from Bent Horn in the Arctic Islands.

The most productive offshore development to date came online in 1997 when oil began flowing out of the massive Hibernia oilfield 350 kilometres (217 miles) out at sea over the Grand Banks off Newfoundland. Taller than the Calgary Tower and half the height of the Empire State Building, the 224-metre- (735-foot-) high Hibernia platform contains drilling equipment, processing facilities to separate water and gas from the oil, a storage area for 1.3 million barrels of oil, and housing and recreational facilities for a staff of 185 people. It's like a village out at sea that produces up to 230,000 barrels of oil per day.

In the early years of the new century these offshore projects provide new production, but none of them can delay the inevitable decline in conventional oil production in Canada. In 2006, for example, crews drilled a record number of gas wells in

The Hibernia platform stands 224 metres high, 33 metres taller than the Calgary Tower (191 metres). (Courtesy Hibernia.ca)

Canada (just over 22,000), but gas production declined because the amount of gas per well continued to drop. Oil wells are also less and less productive, a trend that began in the 1950s. And so the quest for new supplies continues.

New Frontiers

As of 2006, an old frontier and several new ones promise to maintain the Canadian petroleum industry as a major supplier to North American markets. The North has long been the Holy Grail for the Canadian oil industry, but its development has taken time.

An attempt in the 1970s to build an oil pipeline along the Mackenzie River met with opposition from Native and environmental groups, so the government of the day appointed a commission to study the project. The resulting report of the Berger Mackenzie Valley Pipeline Inquiry in 1977 recommended a ten-year moratorium on development until Native land claims and other issues could be resolved.

Creativity may be the final frontier.

With the discovery of additional oil and gas in the North as well as oil prices in the $70 per barrel range in 2005, the project is once again under consideration (was estimated at $7 billion, then $10 billion, then 12 billion and rising). But it is not assured.

What They Do...
Oil Patch Workers: Other careers in the oil patch...

A hundred years ago a job in the "patch" meant adventurous work, hard work, smelly work, and work that seems glamorous to us today. Men built wooden rigs by hand and worked with ropes. Blacksmiths pounded the drill bits out of large chunks of steel. Workers screwed pipe together by hand and coated it with gunnysack material and oil before lowering the pipeline into the hand-dug trench by brute force. At refineries the workers believed they could avoid the effects of deadly sour gas by just holding their breath or leaning away from the leaking valve. Those "good old days" were dangerous, and we are glad they only live on in history books.

Today's oil patch jobs include commercial diver, well logger, chemist, truck and bus driver, financial analyst, surveyor, hydrologist, hazardous waste management technologist, health and safety officer, arbitrator, archaeologist, chef, cook, ecologist, firefighter, insulator, lawyer, librarian, mediator, microbiologist, physicist, plumber, security guard, sprinkler installer, water-well driller, writer, receptionist, statistician, traffic manager, and security guard. The rugged appeal of the old-fashioned jobs may be a thing of the past, but work in the oil patch is now safer, and more people consider it a reliable career. ■

Government, Native, and industry partners quarrel over owner-ship of the pipeline and how to share the costs and profits from the massive endeavour. Even with strong oil and gas prices, it's a big gamble.

Less risky are the huge oil sands deposits in northern Alberta. Though they have been in production since 1967, it costs much more to extract petroleum from the oil sands than from conventional fields. The older oil and gas deposits have run dry, and newer ones are becoming harder to find. In 2001, Alberta produced more oil from bitumen (271 million barrels) than from conventional crude production (264 million barrels). Petroleum from the oil sands also requires huge quantities of water and natural gas in the processing stage.

Unconventional sources of gas also offer promises and chal-lenges. Coal bed methane (CBM), a naturally occurring gas in coal seams, is similar to natural gas. In 2002 EnCana Corporation announced plans for the first commercial coal bed methane proj-ect in Canada. This newly appreciated resource is yet another frontier because of concerns over the methods of producing it and its effects on both the surface and other underground resources. Drilling for coal bed methane is typically much more intensive than for other sources of oil and gas, resulting in a footprint on the landscape that is more of a concern to landowners, who are usually not the same people who own the mineral rights to develop the methane. Surface rights owners also worry about the effects of the drilling programs on their water wells and the water table, or the underground rivers that provide life-giving moisture to people, plants, and animals. As of 2007, almost eight thousand coal bed methane wells have been drilled in Alberta, and the ERCB is holding hearings in affected communities in order to under-stand the concerns of the people most often bothered by the operations of these wells and pipelines.

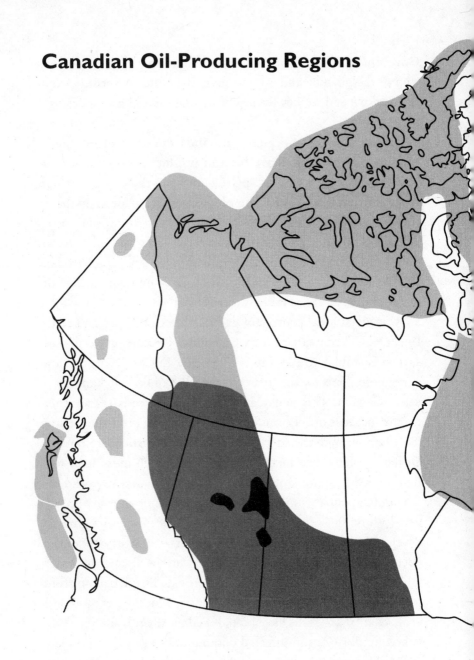

Canadian Oil-Producing Regions

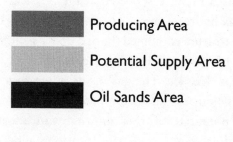

Producing Area

Potential Supply Area

Oil Sands Area

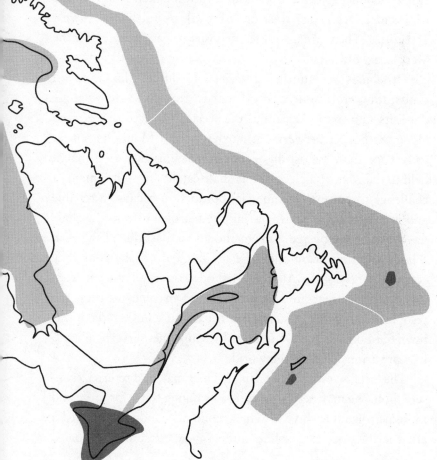

Scientists have also known about another source of natural gas for decades: **gas hydrates**. Naturally occurring like the other hydrocarbons we have discovered at the physical and scientific frontiers, gas hydrates may contain as much fuel as all other sources of petroleum combined. Buried under sediments offshore, they consist of a crystalline solid, a cage of water around a gas crystal, almost like ice. The gas inside is methane. In 2002 an international industry and government consortium began investigating the potential of this unusual source of gas in an experimental project at a remote location called Malik in the Mackenzie Delta, just west of Tuktoyaktuk in the Northwest Territories. There are also huge deposits of gas hydrates off the West Coast of Canada.

Oil shales are another untapped resource. Similar to oil sands, they are a source of oil that is trapped in shale. Huge deposits appear in Australia, the western United States, and along the border between Saskatchewan and Manitoba. Easier to test for than almost any other petroleum source—a match held to the edge of a piece of this shale catches fire—it remains a challenge to put into commercial production because a reliable and efficient means of separating the petroleum from the shale has yet to be discovered. One method is to roast the oil shales in a large rotating drum and take off the hydrocarbons as they come out of the shale. Another method extracts oil and gas from the shales underground, a process that involves heating the resource for months or years. Neither is in use in Canada at this time, but both are in development by companies with shale holdings in various parts of the world.

The last and most challenging frontier may be a return to conservation. We may need to control our raging thirst for more and more petroleum each year. When economic conditions created a crisis in the price of gasoline in the 1970s, consumers cut back.

Perhaps only dramatically higher petroleum prices will motivate and spark the ingenuity necessary to solve our thirst for oil. Is the last frontier our own insatiable demand for petroleum? ■

"Burning ice." Methane, released by heating the ice, burns.
(Courtesy United States Geological Survey)

Key Dates

1863 – Early Canadian oilman drowned in his own oil

1897 – Gas well on the Athabasca River blew wild for twenty-one years

1914 – 1940s – Most gas produced in Turner Valley was flared

1948 – Atlantic No. 3 blowout spilled one million barrels of oil

1957 – Five kilometres of pipeline blew up in Ontario

1970 – Oil barge sank off Prince Edward Island and leaked oil until 1996

1982 – Sour gas blowout in Alberta smelled up the air for weeks

1982 – Eighty-four died in the sinking of the Ocean Ranger drilling platform

1989 – 215,000 barrels of oil leaked out of the *Exxon Valdez*

OOPS!
Mistakes and Lessons

The story of petroleum is also the story of mistakes: oil spills, blowouts, deadly accidents, and carelessness. Industry often learned from its mistakes, but sometimes regulations had to be forced on it by government agencies. The public was always part of this process too, sometimes encouraging the wasteful habits, at other times calling for change. Today's environmental standards are much higher, so we must examine the events of the past carefully, setting them in context of the times. Yet sometimes progress is still the only goal, and despite greater environmental awareness on the part of both the public and industry, a little damage is considered an acceptable price to pay for wealth, power, and prosperity.

Oil Spills and Blowouts

The hazards of oil and gas are not just a nuisance: they can be deadly. As luck would have it, Canada's first oil millionaire, Hugh Shaw, died on February 11, 1863, in one of his wells, the victim of his own fortune. Trying to put out a leak in the tubing that brought oil out of the well, he had himself lowered into the pit in a basket. But the gases in the well started to overcome him, and as his workers rushed to pull the basket out of the well he fell out of it and into the black gold. His obituary in the Cooksville paper read:

> His death was occasioned by suffocation from inhaling obnoxious gases while in an oil well, into which he had descended for the purpose of pulling up a piece of gas pipe. Was within about fifteen feet of the surface; was heard to be breathing heavily, when he fell back into the oil, and disappeared.

The first Canadian oil well was simultaneously a success and a tragedy, and it also created our first oil spill.

Economic conservation was the only concern in the early days.

Petroleum was not only dangerous, its waste was an economic concern. In 1870 Professor Alexander Winchell of the University of Michigan wrote about an oil spill in Ontario in his book *Sketches of Creation*:

It floated on the water of Black Creek to the depth of six inches, and formed a film upon the surface of Lake Erie. At length the stream of oil became ignited, and the column of flame raged down the windings of the creek in a style of such fearful grandeur as to admonish the Canadian squatter of the danger, no less than the in-utility and wastefulness of his oleaginous pastimes.

Winchell estimated the waste at up to five million barrels and called it a "national fortune, totally wasted."

Far away, on the Athabasca River in today's Alberta, the dominion government drilled a well in 1897 at Pelican Rapids. It was a great natural gas discovery, but the well got away on the drillers, and it blew wild and on fire for a generation. Finally, another group of government oilmen snuffed out the blaze and sealed off the runaway well twenty-one years later, in 1918.

In 1901 the Rocky Mountain Development Co. Ltd., based in Calgary, started drilling for oil up the Cameron Valley in today's Waterton Lakes National Park in Alberta. The 1902 drillers met with success, and the well at the Original Discovery No.1 site produced 300 barrels a day of high-grade oil. Another well, owned by Western Coal and Oil Company, on Seepage Creek, blew out and ran wild for two days, pouring oil into Cameron Creek. Fish and ducks died in Waterton Lakes and the nearby river system, and for several weeks the oil film still lay on top of the waters of the Oldman River where it went through Lethbridge, 97 kilometres (60 miles) downstream.

If You Can't Use It, Burn It

Though oil spills are not acceptable by today's standards, it was leaking gas that attracted the attention of early settlers at Turner Valley, in southwestern Alberta. As the story goes, cowboys stopped and cooked bacon and eggs over gas seepages along Sheep Creek. Bill Herron, an Okotoks rancher, noticed gas seepages along the creek and purchased land and formed an oil company. His first drilling adventure was lucky when the Calgary Petroleum Products No.1 well hit wet gas on May 14, 1914. He then built a crude gas-processing plant to take the liquids off the gas, which he sold as gasoline. In the days before pipelines to consumers, the natural gas was virtually useless. The drillers used it for cooking food and heating buildings, but mostly they flared the excess or "waste" gas in order to be able to produce the oil.

Flaring "waste" gas made sense at the time.

The problem of waste natural gas continued for generations. The next major discovery in Turner Valley, at the Royalite No. 4 well in 1924, only made the problem worse. Gas from the 1914 well was relatively sweet, but the 1924 discovery tapped into a reservoir of highly pressurized sour gas, and the well blew wild. For weeks it burned, consuming the rig and anything else nearby with its massive pillar of flame. Attempts to extinguish it with steam failed. Finally, two Oklahoma experts snuffed out the flame with dynamite—the explosion of the dynamite robbed the flare of oxygen. The drillers then diverted the gas flow and capped the well. While it was a blowout, it expelled gas into the atmosphere just southwest of Calgary at the rate of 24 million cubic feet of naphtha-rich gas per day. Even after it was harnessed, most of the gas from this well was burned in a nearby coulee called Hell's Half Acre. Dozens of other wells sprang up after this discovery, and most of the gas from them also burned in flares.

Gas flares at Home Oil well, Turner Valley, Alberta circa 1940s.
(Glenbow Archives, NA-4062-6)

Hell's Half Acre

For decades Turner Valley's flares turned night into day, but the most famous of the flares was Hell's Half Acre. Located in a streambed coulee, or valley, just northeast of the town of Turner Valley, it rumbled and shook and roared. Waste natural gas—sour gas that was deadly due to high concentrations of hydrogen sulphide—screamed out of the end of several pipes and down into the coulee. A brave worker tied a rock up into a rag, soaked it in gasoline, and then lit it on fire and threw it into the valley to light the gas. The explosion rocked through the community.

Residents of the town said the ground shook all the time, a low rumble. Plates in the cupboards rattled, and the noise made it hard to hear anyone talking if you were near the flare. The burning sulphur smelled like rotten eggs, or as western Canadian author W. O. Mitchell called it, "A big, stinky old fart!" ■

Flares burning "waste" natural gas into a coulee just north of Turner Valley town-site, 1926. (Glenbow Archives, NA-1716-5)

The 1936 discovery of crude oil by the Turner Valley Royalties No. 1 well in the middle of the Turner Valley oilfield created the third and longest boom for this oilfield. The "sensational" discovery also "scattered crude oil over a wide area in the vicinity of the derrick." The 1914 and 1924 discovery wells produced wet gas, but the 1936 well hit crude oil. Black gold. Texas tea. Coming in the middle of the Depression, the oil strike created jobs and helped the residents of southern Alberta forget the hard times. Prosperity had arrived yet again. As for the waste of gas and some of the oil, who cared? There was lots more to go around, and this time it seemed the boom would last forever!

"Sour gas" was the smell of money.

Gas waste in Turner Valley was not a trivial matter, and many people complained about the practice. In 1938 the *Daily Oil Bulletin* in Calgary noted that 75 wet gas wells were wasting about 200 million cubic feet of gas each day in order to recover about 1,800 barrels of naphtha. In 1939 a *Maclean's* magazine article stated that 35,000 cubic feet of Turner Valley gas were being burned off in order to produce every barrel of oil. At 1939 prices this meant that $10 of gas was being wasted in order to produce $1.20 of oil. Bill Knode, hired by the Alberta government to take charge of the waste problem under the new **Petroleum and Natural Gas Conservation Board** (today's Energy Resource Conservation Board), said, "This is a crazy set up. If you let this gas get away, how are you going to raise [lift] your oil? And if you can't raise your oil, where will you be?"

Not only was gas flaring bad for the long-term viability of the oilfield, the waste defied common sense. Residents of southern Alberta were giving up coal and wood in their furnaces and stoves in favour of natural gas. The 200 million cubic feet of "nuisance" gas that was being flared in 1939 was enough to heat thousands of houses every year. But economics triumphed over common sense; with no pipelines from most of these wells to limited markets, the owners of the wells had to flare the gas in order to produce the liquids. The Canadian Western Natural Gas Company had a monopoly on the sale of natural gas in southern Alberta and had more gas than it could use much of the year. It stored some in underground gas reservoirs during the summer, but most of the gas produced in Turner Valley did not get to market.

The Petroleum and Natural Gas Conservation Board

As a result of the mistakes made in the Turner Valley oilfield, the Alberta government set up the Petroleum and Natural Gas Conservation Board on July 1, 1938. The new board's powers

Hydrogen sulphide was dumped from the top of the two towers at the right from 1925 until 1952 and allowed to mix with fresh air at Turner Valley Gas Plant, circa 1930s. (Glenbow Archives, NA-67-53)

were wide ranging, but its biggest concern was economic waste of the petroleum resource. It set out to control the waste flaring, and on August 13, 1938, increased well spacing from 6.7 to 13.4 hectares (20 to 40 acres) to prevent overcrowding of wells in the Turner Valley field.

The "rule of capture" makes it legal to steal petroleum from your neighbour.

Overcrowding was another concern. Gas and oil are trapped in giant reservoirs deep below the surface and migrate to areas of lower pressure—just like water runs downhill. If one company produces petroleum quickly from its well, it can create a lower pressure area in its part of the oilfield and actually suck product out from its neighbour's property. First come, first served—the **"rule of capture"** meant that the first person to get the oil out of the ground could claim it. The Conservation Board tried to promote fairness by preventing overcrowding of wells—no more than one every 13.4 hectares (40 acres)—and limiting flaring in order to prevent the rapid depletion of the pressure in the field.

The engineers at the Conservation Board worked hard to enforce the laws that prevented economic waste. Today's definition of "conservation" is quite different from the one used in the 1930s—we think of the environment when we talk of conservation. Back in the 1930s the Conservation Board field inspectors had their hands full just trying to prevent the waste of gas and oil, and so the 1948 blowout and massive oil spill at Atlantic No. 3 were a challenge to the officials at the Conservation Board.

Hugh Leiper was working on the rig that infamous day, March 8, 1948:

I had a cheap little Brownie camera and after getting off the tower at 8:00 a.m. I went back to our camp, had breakfast and came back and took this picture. You can see where I took it from standing on the north/south road and facing east to the rig. Also you can make out the missing sheets of tin which we knocked off the pump house immediately after the blow-out using 2" by 4" and 2" by 6" boards. In addition, please note the position of the outdoor toilet where a few days later Cliff Covery lit up a cigarette causing a fire which spread to the banks of the sump. As you know we had an extremely difficult time in putting out the fire and also preventing it from spreading inside of the dyke walls which would have ignited the whole area. Putting out this fire with having only meager material was a miracle. When I think of what the consequences could have been for us I can only shudder. I guess you blame it on youth and ignorance. In any event I had the small snapshot blown up and it turned out extremely well.

Residents of Edmonton tasted oil in their drinking water.

The 1948 blowout happened just a year after the famous Leduc discovery. A Scottish farmer near the Leduc well, John Rebus, sold his 53.5 hectares (160 acres) of land to Frank McMahon for $200,000 cash and the usual 12.5 per cent royalty. McMahon's first two wells on the lease were good, but the third was a blowout of epic proportions. Atlantic No. 3 well erupted into Canadian petroleum history in early March 1948, when it blew in wild. Over the next six months the well produced more than a million barrels of oil and in excess of 10 billion cubic feet of gas. For the last three days of its unfettered career, the gas was on fire. Crews drilled relief wells on either side of the blowout and eventually succeeded in stopping it.

In the end, the Conservation Board took over the disaster and managed it as best as possible given the circumstances. **Berms** around the field kept in most of the oil, but some escaped into the North Saskatchewan River, resulting in polluted drinking water downstream, including the Edmonton water supply. Workers eventually drilled two relief wells to control the gushing oil and then cleaned up the site according to standards that were appropriate for the times. No crops grow on that land to this day.

They stopped the blowout, cleaned up the oil, and shipped it off to the refinery. That's all there was to it. The expense of controlling the well came from the sale of the oil that was recovered from the area. Little thought was given to the environmental consequences of this oil spill as the boom in the oil industry continued into the 1950s. Reporter Les Rowland recalled the blowout at Atlantic No. 3 as "more of a sensation than a problem."

Pipeline Problems

A busy industry in the late 1940s needed markets, so oil companies began making plans to build the first of many pipelines from the prairies to distant markets. In 1950 Interprovincial Pipe Line Company built the world's longest pipeline from Edmonton to Superior, Wisconsin, in just 150 days. Typical of the times, the company built quickly, efficiently, and with a single-minded purpose that did not always keep in mind the needs of others. According to the history of the company, most of the 2,100 landowners through whose land the pipeline passed were co-operative. There were only forty-one expropriations, and only three were considered "unco-operative."

One of the men involved in getting permission from landowners said, "It was really very exciting for us. We were on our rollerskates and there wasn't much wind against us. It was just a hell of a time." Opposition to the pipeline was minimal, but in the years that followed it became evident that the company was still learning the importance of dealing with landowners and taking care of the environment. In their rush to build the pipelines, companies often overlooked the consequences of tearing up farmers' fields along the pipeline route and the destruction of aquifers (underground rivers).

Early pipelines created a mess.

Even the federal regulators—the National Energy Board (NEB)—did not show much concern for these matters in the early days. As a consequence, a company made a mess out of some farmland in Ontario and refused to clean it up. The company's lawyer said in a court case, "Right or wrong, a pipeline company can go into a property and turn it into a wasteland." The affected farmers had to fight through the courts for years before the regulators and the companies implemented more careful procedures and considerate policies. But at the time, getting the pipeline built was of utmost importance to the government and the oil industry, and the public raised no alarm over the pipeline's construction practices. The landowners, however, forced changes to the way the NEB and the pipeline company conducted their operations in the future.

Though pipelines are usually a reliable and efficient way of moving petroleum, there have been some spectacular accidents. An explosion on the TransCanada gas pipeline on Christmas Eve of 1957 near Dryden, Ontario, created "a tremendous roar, the ditch heaved, the cover of soil and rock flew upwards, and the pilot of a Trans Canada Airlines (now called Air Canada) airplane saw a long flash of light leap from the earth below him. The longest pipeline break in history, some three and a half miles, took place instantly." The pipeline that burst met all prevailing standards, so its regulator, the NEB, began working with the Canadian Standards Association to create pipelines that would better meet the demands of the harsh Canadian environment.

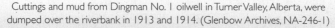

Cuttings and mud from Dingman No. 1 oilwell in Turner Valley, Alberta, were dumped over the riverbank in 1913 and 1914. (Glenbow Archives, NA-246-1)

Offshore Disasters

Pipeline leaks are easier to contain than those at sea. In January of 1969, for example, 6,000 barrels of oil leaked from an offshore well and fouled the beaches for 48 kilometres (30 miles) along the coast of Santa Barbara. When the people of California faced the peril of pollution on the beaches of their "backyard," they reacted quickly, and the state declared a moratorium on drilling.

In Canadian waters, the oil tanker *Arrow* ran aground and spilled its oil into a lobster fishery off Nova Scotia in 1970. This accident caused the public to question the development of offshore oilfields and fear that similar accidents could become common if these new areas were allowed to develop.

Then on September 7, 1970, the oil barge, the *Irving Whale*, sank 60 kilometres (37 miles) off Prince Edward Island in the Gulf of St. Lawrence. A 400-square-kilometre (154-square-mile) oil spill developed during the next two days and fouled 80 kilometres (50 miles) of shoreline on the Magdalene Islands. Over the next twenty-five years a quarter of the 4,200 tons of oil and some PCBs leaked out of the old tanker. Finally, in 1996, the barge, as large as a football field, was refloated, and its tanks were cleaned.

In Fear of Spills

Fearing oil spills, the British Columbia government banned oil exploration in the Georgia Strait between Vancouver Island and the British Columbia mainland effective May 1, 1970, and the federal government in Ottawa banned exploration on the entire West Coast in 1971. ■

In 1972 Canadian concern about the tanker traffic along the West Coast was high and probably contributed to the American decision to build the Trans-Alaska Pipeline System (TAPS) between 1974 and 1977. An accident involving a tanker at Cherry Point, Washington, confirmed the fears of a crude spill. Although the spill from a broken valve was quickly stopped, about 45,000 litres (12,000 gallons) of crude oil drifted into Canadian waters, fouling the coast. The Canadian ambassador pressed for a cleanup and full compensation of any costs.

Canadians had every right to be concerned. Thousands of oil tankers run up and down the West Coast through Canadian waters, and an oil spill on the high seas can be a disaster. But Canadians were not in control of the Pacific shores of North America, and American plans to exploit the oil in Alaska presented a real threat. The next big oil spill could easily have happened in Canadian waters.

The *Exxon Valdez*: A Case Study in Lessons Learned

Canadian and American governments feared more oil spills along the West Coast, yet oil from the port of Valdez, Alaska, had to find a way to market. And so the American government created an extensive system of regulations and safeguards to protect its waters from disaster. Hearings and delays infuriated the old-style oilmen, but safety was important. Double hulls were promised for this tanker route when the American government gave its permission. However, the industry fought the costly new expenditures, arguing that they would make moving the oil prohibitively expensive. Although this requirement was

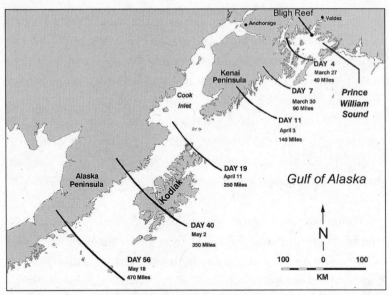

Exxon Valdez spilltrack. (Courtesy David Page)

relaxed, 27 per cent of the oil tankers used to transport oil out of Valdez were double hulled. Experts disagree about the safety value of double-bottomed, or double-hulled, tankers to this day.

Once oil started flowing through TAPS in 1977, up to seventy-five tankers per month arrived at Valdez, Alaska, where they filled up with oil from the Alaska North Shore oilfields and then returned to ports along the American West Coast with their precious cargo. Thousands of tankers came and went silently up and down the coast of western Canada and the United States without causing much concern during the 1970s and much of the 1980s.

A disaster was inevitable.

One alternative to TAPS and the tankers moving oil from Alaska to the refineries farther south could have been a pipeline across Alaska and then through the Canadian North. In 1971

David Anderson, chairman of the Commons Committee on Environmental Pollution, disapproved of TAPS, saying that "it's inevitable that some disaster will eventually happen" if tankers were part of the scheme. British Columbia government officials also stated their strong opposition to any tanker traffic along the Pacific coast even though Canadian jurisdiction was limited to the 19-kilometre (12-mile) limit.

Plans for the Canadian route, the Mackenzie Valley Pipeline, began in earnest in 1971 as Canadians tried to show that it would be safer than the one across Alaska which passed through earthquake territory and then in tankers down the West Coast. The studies also investigated the environmental and social consequences of a pipeline. Panarctic found considerable quantities of gas in the Arctic that year, and the prospect for oil and gas in great quantities in the North looked better than ever before. But the Canadian government set up hearings into the pipeline, and on May 4, 1977, the Berger Inquiry into the Mackenzie Valley Pipeline released the first part of its report. The recommendations called for a ten-year moratorium on Mackenzie Valley pipeline development to allow time to deal with environmental issues as well as the land claims of the people of the North. Within months, Gulf Oil Canada and Mobil Oil Canada ceased drilling in the Mackenzie Delta area of the Arctic until the pipeline debate was resolved.

Unwilling to wait for the Canadian government decision, the Americans started building TAPS in 1974, and oil began flowing through it in 1977 to the port of Valdez, Alaska. There, the first of more than 8,700 tankerloads of oil left for the southern states without a mishap. Art Davidson's book *In the Wake of the* Exxon Valdez: *The Devastating Impact of the Alaska Oil Spill* examines the tanker issue before and after the big oil spill. Although many opposed the tankers during the public debate before the line was built, the wealth brought by the line soon made most Alaskans believers in dependence on oil. One citizen is quoted in Davidson's book:

> Like other Alaskans, I soon became accustomed to the benefits of having the oil industry in the state. With the advent of North Slope oil money, state income taxes were abolished, and each resident began receiving an annual windfall check of at least $800 from the state's oil revenues. Oil development was soon producing 85 percent of state revenues, creating jobs, and paying for new highways, schools, libraries, and performing arts centres. For nearly twelve years, we enjoyed the prosperity of our state's wealth without having to face its trade-offs ... until the wreck of the *Exxon Valdez*.

Oil made Alaska fantastically rich.

The good times of the 1970s and the high price of oil ended in the early 1980s. Bad luck came too. Human error was bound to catch up with the tankers that moved up and down the West Coast. As early as 1984 standards for pollution control related to TAPS began to sag. People with knowledge and training in the Alaska Department of Environmental Conservation were gradually moved away from Valdez through budget cuts and transfers. The Alaska politicians only budgeted for 10 per cent of what the department said it needed to keep track of the pipeline operation. Once the pipeline company noticed that the government was not taking things seriously, it too cut back on its commitment to being prepared for an oil spill.

Eventually a tanker, the *Exxon Valdez*, ran aground on a reef near Valdez in 1989, and every North American sat up and took notice. The resulting oil spill spread 215,000 barrels of oil over the waters of one of the most beautiful coastlines anywhere. It killed fish, birds, mammals, and many other forms of life.

When the accident happened, it took almost everyone by surprise. How could this have happened? According to one account of the spill and its aftermath by Alaska resident Art Davidson:

Molusk covered with oil. (Courtesy *Exxon Valdez* Oil Spill Trustee Council)

Reports began to surface of watered-down regulations, oil company budget cuts, Coast Guard cutbacks, rule violations, alcohol abuse, fatigued tanker crews, and government negligence, and it soon became clear that many factors besides chance had contributed to the accident's cause.

The environmental safety systems seemed acceptable at the time.

After more than a decade without an accident, everybody thought the system was safe, and that planning for a large spill was a waste of valuable time and money. Davidson believes, "Stated simply, environmentally safe oil development and transport were and are impossible: they always entail a degree of risk."

Frank Iarossi, Exxon's first man on the job at Valdez, said of the cleanup effort:

> It's just that it was totally inadequate relative to the magnitude of the spill. I'd say the lesson to society is that a spill like this can happen: no matter how low the probability, the potential is still there for it to happen. Another lesson is in the inadequacy of current technology. What we have is just not good enough, no matter how finely tuned an organizational structure you have.

No one could have handled the oil spill that spread out from the *Exxon Valdez*. Congressman George Miller of California said of the spill:

We have a world-class demonstration of failure
... I don't point a finger just at Exxon, because
you have assembled everything available in the
world. Your competitors would have had the
same problem.

How can the public feel comfortable with
how you would handle a spill off the coast of
California, South Carolina, Florida, or any other
coastal state? If the industry can't respond to this
spill on a timely and effective basis, I don't think
the public is going to buy into new offshore development ...

Unless some drastic changes take place, I just
can't ask my constituents and the constituents of
other states to run the risk of this kind of disaster.

As people began understanding the consequences of an oil
spill in today's environmentally conscious society, they squirmed
with psychological and emotional pain. Valdez councilman
Lynn Chrystal said, "Valdez has always been one of the most
pro-development communities in Alaska ... But we feel let
down and very cheated." The people of Valdez thought of the
pipeline company as a good member of the community until the
spill occurred, and the company, was totally unprepared to
clean it up. Few tried to check up on the company and those
that did were thwarted by the size and financial power of the
huge corporation.

The oil spill felt like a betrayal.

A long-time resident of Alaska said, "The thing that is hardest
for a development-minded individual is this tremendous sense of

betrayal. I believed our state officials when they said they were taking care of things. And I believed the oil companies when they said we could have both a pristine environment and oil."

At first everyone worked on the assumption that technology existed to undo the damage caused by the oil. With time, the very concept of "clean" became subjective. At first Exxon said it would clean everything up, all the oil. That was impossible logistically and economically. Later it referred to the beaches as "treated" and left nature to do the rest. Even though Exxon spent billions of dollars on the operation, it is evident that no amount of work could have completely cleaned up the oil. While some locals felt that the company did not do everything it could, others used the spill as a chance to make enormous and

Booms cleaning up oil spill. (Courtesy *Exxon Valdez* Oil Spill Trustee Council)

outrageous profits by leasing their equipment to the oil cleanup effort. Although there was some co-operation at first, by the end of the process there were many bitter feelings on all sides.

Many stories of individual action come out of such a catastrophe. Dr. Jim Scott treated some of the bald eagles, America's national symbol, that were affected by the spill. His work made a difference. According to Davidson,

> If this energy can be directed at prevention inside as well as outside the industry, we might reduce the impact of major oil spills and begin solving some of the larger environmental problems of which oil spills are symptomatic. For me, one enduring lesson of the *Exxon Valdez* spill is how direct the line is from individual conscience to global problem solving.

Locals were not the only ones who felt pain as the result of the spill and took action. In the days after the spill, many Exxon credit card holders cut up their cards and sent them to the company. They marched outside the company's annual meeting, and some demanded that CEO Lawrence Rawl resign. Pension fund representatives with investments in the company gave the company notice that its actions were under close scrutiny.

Finally, a man in charge of a park in Alaska put the spill in perspective. Ray Bane, superintendent of Katmai and Aniakchak national parks, some 645 kilometres (400 miles) from the spill and whose beaches were fouled by the oil, said, "All this devastation, that should not have happened. I think we all fell asleep at the wheel, myself included. As a citizen, I have a responsibility to put pressure on state and federal officials so something like this will never happen again."

Other Lessons from the 1980s

Hindsight is always a useful tool. Another incident turned into a disaster in 1982 when the Ocean Ranger drilling platform sank in a storm off the East Coast, killing all eighty-four crew members. As the worst marine disaster in forty years, the tragedy was front-page news. While the commission into this incident carried out its investigation, Newfoundland barred winter drilling due to fears for the safety of drilling crews and the inability of Ottawa to provide rescue in case of an emergency. The Ocean Ranger report stated that the drilling platform sank due to errors by most of the responsible people. In 1985 the commission suggested major changes to safety regulations, rescue systems, and weather forecasting. The extremely hazardous nature of the work, the complicated equipment required to operate under the severe conditions, as well as the ever-present danger of human error caused unfortunate loss of life. More than six hundred people have died on drilling platforms around the world since 1965, but immersion suits, covered lifeboats, and helicopters, as well as more safety training, are helping to keep loss of life to a minimum.

We learn from our mistakes or we repeat them.

In Alberta in late 1982 a sour gas well blowout at Lodgepole captured the attention of Albertans for weeks. The nightly news covered the details of one man's death and the fifteen who were knocked unconscious while trying to cap the well. The workers finally brought the well under control on December 24, 1982.

Well blowouts happen more frequently than anyone wants, but this one received massive media coverage. It attracted interest because people had to be evacuated from the area, and the smell of the sour gas reached major metropolitan centres throughout Alberta and Saskatchewan. An inquiry into the incident was made public in late 1984. It suggested that, theoretically, the company could have avoided the sour gas well blowout with better training and equipment.

Lessons Learned and Lessons to Learn

Mistakes happen all the time, especially during boom times in the Canadian oil patch, but as each of the stories in this chapter shows, mistakes teach us how to be more careful and take better care of the environment.

But we have also let things slip in some areas, including our overall consumption of oil and gas—we use much more now than we did thirty years ago. Government regulators allowed automobile efficiency standards to lapse by making them voluntary in the late 1980s, and the deregulation of the oil patch that same decade has relaxed some environmental standards too. Allowing companies to self-regulate may be more efficient, but laws without enforcement are easier to violate.

The challenges remain. We need to use energy wisely and to take responsibility for the mistakes of the past. We need to balance our need for oil against the damaging effects of burning fossil fuels. Global warming is now a fact, though heated discussion continues over what to do about the ways humans contribute to climate change.

Yes, we goofed. We have learned from our mistakes, but we are also making new ones. Diligence is as important now as at any other time in the history of the oil patch. ∎

PLEASE GOD LET THERE BE ANOTHER OIL BOOM.
I PROMISE NOT TO PISS IT ALL AWAY THIS TIME.

The words from a famous 1980s Alberta bumper sticker.

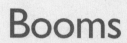

Booms
This Time It's Different

On November 22, 1950, a gusher spewed forth from under the earth at John A. Gow's farm near Barons, Alberta. When someone asked him if he had seen the gusher, he replied, "Not yet. No point in getting excited. I've waited 36 years for this day, and I reckon the oil will keep for another hour or two until I've had my lunch."

**Booms and busts are a normal cycle
for resource industries.**

Each time we think, "This boom will last! The boom and bust cycles are over. This time the conditions have changed, and the boom is permanent. It will last forever." But the booms don't last forever, and people get stung. Disappointment settles on the resource development sector of the economy, and people move back home. Hope gets shattered again, and investments disappear overnight. Just when it seems like the economy will never recover, the price of oil starts climbing again.

Demand

Many factors influence the boom and bust cycles in the Canadian oil patch. The first is demand—a fickle master indeed. In the early days crude oil found a market by replacing other sources of oil. Until the 1850s, most oil used in Canadian homes came from whales. It burned in lamps, appeared in some cosmetics, soap, paint products, and other consumer items. When the population of whales dropped because we hunted them almost into extinction, the whale oil scarcity prompted Abraham Gesner to create kerosene. Made from coal gas, its original name was "**keroselain**," derived from the Greek word for "wax oil." Today's kerosene comes from crude oil; it's heavier than gasoline, but a similar product.

Coal oil, as it became known, was the first petroleum product to really take off. It quickly replaced whale oil, and by 1860 about seventy plants were making oil from coal throughout central Canada and the eastern United States.

*Petroleum was a local industry in 1900
and international by 1950.*

Petroleum also became a substitute for two other fuels: wood and coal. Most homes used wood for heat and cooking, and coal provided a more concentrated form of heat. Then came cars, trucks, airplanes, tanks, and many other vehicles that used internal combustion engines. Demand grew quickly during the first half of the 1900s and even faster after World War II. Oil also became useful to industry, and the plastics revolution took even more petroleum. Construction materials for homes, medicines, toys, and numerous other consumer items also began to require petroleum.

As the demand for oil increased, so did the incentives for finding it anywhere in the world. Exploration for other minerals often identified petroleum resources long before there were markets for products. However, motorized vehicles moved closer to the remote oilfields, and local markets arose. The Turner Valley oilfield in southern Alberta, for example, supplied raw gasoline to car owners from before World War I to the 1940s. But with the discovery of vast amounts of oil at Leduc in 1947, companies needed access to larger regional markets. Soon, they required access to the whole continent.

With access to markets, the Canadian petroleum industry pulled out all the stops and began exploring and producing at maximum capacity. As a result, the industry's exploration rate peaked in the mid-1950s: that is, the amount of gas and oil found per well began to drop less than a decade after the spectacular discovery of oil in the Devonian formations near Edmonton. This fact was unimportant for a time, as long as companies were able to find new fields and drill more and more wells each year in order to keep adding to reserves. However, companies had to spend more time, energy, and cash to find these more expensive supplies. When the price of oil dropped, and it always drops again, the expensive oilfields are too costly. Exploration shuts down, and reserves dwindle even though production continues apace. When the next boom comes along the exploration and drilling programs take off again, and the cycle continues.

Elephants, Old and New

When exploration fails to find new **elephants**—the big oilfields—companies return to old ones with new recovery techniques that might include **water-flooding**, literally pushing water into the reservoir to increase its pressure and move oil to the wells. Companies also rework the wells and develop new drilling strategies to liberate more product from the reservoir, or they just drill more wells, deeper or horizontally, whatever is required. In fact, about 90 per cent of the oil remains in the ground after a field is fully drained, so future generations of engineers may find new ways to recover even more from abandoned fields.

What They Do...
Economists and Accountants: Complex and Volatile Forces

Energy economists predict the future, petroleum accountants uncover the past.

In the 1920s fat cats in suits smoked big cigars and drank whisky in the Palliser Hotel in Calgary as they dreamed up oil companies. Some drilled for oil, others just drilled for the cash in the pockets of naïve investors. The energy economist was still to be invented.

But today's oil patch is a business that must make money for shareholders, so each company works hard to keep track of the financial side of the operations. Over the course of a decade, the international price of oil may ping-pong back and forth between $5 and $75 many times. It's the job of the energy economist to understand the many complex and volatile forces that affect the price of oil. These business analysts create strategies that enable oil company officials to make wise decisions.

Some people may not think that accountants are as sexy as economists, perhaps, but their work is equally important. They keep track of budgets, costs, changes in inventory and pricing, and hundreds of other details. At the end of each year they prepare the financial documents that show a profit or a loss—and a profit is what investors look for when shopping the stock market. ■

*Supplies, markets, and transportation all
affect the price of oil.*

After conventional oilfields in accessible regions expire, the search goes to the frontier—both geographic and technological. The oil sands became economically viable in the 1960s, and Canadian natural gas that was burned as waste for decades found a market when sent to distant consumers through pipelines. Offshore drilling holds promise today, as does the exploitation of coal bed methane, gas hydrates, and other cutting-edge ventures (see Frontiers—Oilfields on the Edge, pp. 121–35).

The Danger of the Global Marketplace

As soon as the Canadian oil patch expanded beyond regional supplies and customers into the global marketplace, it became much more vulnerable to the world oil price. Canadian oil is expensive to produce compared to Middle Eastern crude, but when the international price of oil is high, companies flock to Canada to produce our petroleum. Then, when the price of oil drops again—and it always does—the international price of oil creates a recession in the Canadian petroleum industry. Regardless of the real cost of producing oil in Canada, companies buy and sell oil based on the international oil price.

*International economic and political events
are unpredictable.*

World wars and other international military events also have a great effect on world oil supplies and on the price of crude. International conflicts that affected the Canadian oil industry included World Wars I and II, the Suez Crisis in the mid-1950s, the Arab-Israeli conflict in the early 1970s, and more recently the two Gulf Wars, in 1990 to 1991 and then again ten years later. As supplies tightened and prices bounced dramatically, our oil patch felt the shocks, both good and bad. This cycle continues, and news stories often blame international tensions for a rise or fall in oil prices.

Foreign Policy

International foreign policy has also affected Canadian oil. Petroleum has been a strategic resource for a century and British and American foreign policy has relied on Canadian oil supplies during times of war. During World War II, for example, the British Commonwealth Air Training Plan relied on Canadian facilities and oil supplies to help it train thousands of aircrews at more than one hundred Canadian flight schools from coast to coast.

Even though Canada was not self-sufficient in petroleum at the time, the United States took a major interest in the Canadian North during World War II. Under its leadership—perhaps even control—engineers built and operated a wartime pipeline called the Canadian American Norman Oil Line (CANOL) between the Northwest Territories and the Yukon to provide gasoline for their war effort.

Given the American reliance on imported oil, Canada was once again called upon to deliver petroleum during the Suez Crisis, the Arab-Israeli war, and more and more in recent decades. But proximity to the United States has not always worked in our favour. When economic realities change we are sometimes the first to suffer. For example, American oilmen in the central states needed access to markets in the late 1950s, so Canadians "voluntarily" limited their exports across the border

The CANOL Trail

CANOL (Canadian American Norman Oil Line), a World War II pipeline, moved oil from Norman Wells on the Mackenzie River in the Canadian Northwest Territories to a refinery in Whitehorse in the Yukon, 950 kilometres (590 miles) away, for a cost of about $138 per barrel.

Under the threat of a Japanese invasion, the pipeline was pushed through quickly, and the pipe just sat on top of the ground. It crossed rivers and went through a mountain range, moving oil in aid of the war effort. Today hikers walk the historic trail, but the river crossings and grizzly bears make the long trek a challenge. ∎

Pumping station number 3, mile 76, CANOL road, Northwest Territories, 1944.
(Glenbow Archives, NA-4450-64)

under pressure from the American government in the 1960s. Yet during the Arab oil embargo in the 1970s the Americans needed all the oil we could supply, just as our National Energy Board predicted that our supplies were running low. The United States has become even more reliant on Canadian oil in recent years because of its ever-increasing appetite for petroleum, and Canada is currently the largest single supplier of crude oil to the American people.

International Markets

International markets for products related to petroleum also affect the Canadian oil patch. For more than twenty years, from 1925 to 1952, the hydrogen sulphide that was stripped from sour gas at the Turner Valley gas plant in Alberta was dumped out of the top of two tall towers to mix with prevailing winds—the solution to pollution was dilution. Then, during World War II, the federal government needed sulphur for armaments, and it subsidized the production of the product for the war effort. After the war the oil companies began selling sulphur to the rubber industry and to chemical and other industrial users. Though it has been a great source of income to the oil patch, its price has fluctuated wildly over the decades, depending on international demand.

Natural gas was another waste product for generations. Early oil companies flared the gas in order to produce oil and gasoline, but in the process greatly reduced their ability to extract oil from the fields in the long term, because without the pressure of the gas in the field the oil would not move to the wells. At one time Turner Valley producers were burning off $10 of natural gas in order to produce $1 of oil. They finally quit this wasteful process when wartime shortages of other fuels created markets for natural gas, propane, butane, and pentane during World War II.

The price of natural gas rose from a few cents to $12 in fifty years.

During the 1950s and 1960s more homes converted to natural gas heat, and during the shortages of oil in the 1970s natural gas became even more popular. Even so, natural gas prices were just a few cents per MCF. Then in the 1970s the price jumped almost 700 per cent in five years to $2.25 per MCF in 1977. The price of gas sailed past $12 per MCF during the cold winter months of 2005–06 and then fell to less than half the peak price. These wild fluctuations are caused by many supply and demand factors that include the weather, pipeline capacity, technological advances, and speculation on the stock market.

Concerns over the environment starting in the 1970s further raised the value of natural gas, when electrical generation companies in the United States converted their plants from dirty coal and fuel oil to cleaner-burning gas. Sales of natural gas have risen dramatically over the past few decades, even though the price for the product varies greatly too. Canadian consumers pay the world price for natural gas, so the cost to the homeowner can be a cause for concern.

International Economy and Interest Rates

Other economic factors that affect the Canadian oil patch are the international economy and interest rates. The downturn in the 1980s was influenced, in part, by a worldwide recession and the rise in interest rates. Given our close economic ties, our economy goes into a tailspin whenever the United States suffers a recession. When interest rates rise at the same time, construction projects become expensive just as the price of oil drops. The result is often deadly for new projects, for individuals, and for the whole province or region.

All these economic forces, and many more, affect each boom and bust cycle. The details of the dramatic bust in the 1980s show just how far we can fall after a lofty ride. Just as surely as the price of oil rose in the 1970s, it fell in the 1980s, from a high of almost $45 in 1980 down to less than $19 in 1988.

Bust: The Dirty Eighties

The bust was particularly hard on the tens of thousands of people who had moved to Alberta to cash in on the boom. Calgary quit building new office towers for the oil companies. People lost their jobs. Housing prices fell by 50 per cent. Some people even quit paying their mortgages and lived rent-free until the banks foreclosed on them after a few months.

As a province, Alberta did little better than many of its citizens. During the boom it had spent wildly, relying on oil income for half the budget. In one three-year period under Premier Peter Lougheed, spending increased 50 per cent. The civil service grew during this boom as did the number of hospitals, universities, museums, hockey rinks, and many other services— facilities that were cheap to build during a boom but expensive to maintain during a downturn.

In early 1982 OPEC cut production to 17.5 million barrels per day in the face of a decreasing demand for oil (down from a peak of 64 million barrels per day in 1980 to 62 million in 1981 and 55 million in 1982), providing another example of the change in the international petroleum picture. Ottawa responded to the dropping price of oil by announcing $2 billion in changes to the National Energy Program provisions, and for its part, Alberta provided an additional $5.4 billion in changes to its royalty and other income programs to provide more incentives for exploration and production in the province. The Lougheed government balanced its budget by reducing the proportion of its income that was to be diverted to the Alberta Heritage Trust Fund from 30 per cent to 15 per cent, thereby postponing the inevitability of a deficit if petroleum revenues did not rise quickly (half its income was from non-renewable resources). In the face of falling oil prices Alberta oil was not

Alberta's Piggy Bank and Provincial Golf Course

The **Alberta Heritage Trust Fund** was created by the Peter Lougheed government in Alberta in 1976 "to save for the future, to strengthen or diversify the economy, and to improve the quality of life of Albertans." During its first thirty years it generated $29 billion in investment income and its value was $15 billion in 2006.

Each year the Alberta government withdraws about a billion dollars from the trust fund for use in current projects. Over the years these have included debt reduction, health care and education initiatives, infrastructure construction, and other capital projects. One of the most famous is the golf course in the Kananaskis—Alberta's own mountain park. ■

"Because it's there!" is the sentiment of all Alberta political parties wanting access to the Alberta Heritage Savings Trust Fund, according to cartoonist Tom Innes in the *Calgary Herald*, October 15, 1982. (Glenbow Archives, M-8000-1129)

competitive on the international marketplace, and companies chose to leave it in the ground rather than sell it at a loss.

The most serious economic downturn in the United States since the 1930s contributed to a corresponding economic recession in Canada, but oil exports to the Americans increased 57 per cent, and gas exports grew about 3 per cent in the face of the ongoing Iran-Iraq war. Drilling fell again, by 17 per cent, to six thousand wells, and Canadian domestic consumption fell overall, with only a slight increase in the use of natural gas. Petroleum imports fell dramatically, by more than 45 per cent.

Rigs that left Alberta in the 1980s were cut up for scrap in Texas.

As the recession seemed to be bottoming out, the Alberta and Ottawa governments signed yet another new pricing agreement in the middle of 1983 that froze the Canadian price of oil just as it fell to less than $40 per barrel on the international market. Though Alberta diverted investment income from the Alberta Heritage Trust Fund to general revenue and held the line on budgetary increases in many areas, it nearly wiped out

Drill rig crews fly south over the head of Marc Lalonde, federal Minister of Energy, Mines and Resources, in this Tom Innes cartoon that appeared January 30, 1981, in the *Calgary Herald*.
(Glenbow Archvies, M-8000-757)

the accumulated surplus in its General Revenue Fund with a deficit of more than $2 billion. Unemployment in the province peaked at over 12 per cent in March 1983 with more than 300,000 people claiming insurance payments, up more than 70 per cent in just one year. Building permits shrank by half, and a full quarter of the workers in the construction industry lost their jobs. It came as no surprise at the end of the year when population records showed a net loss of almost 3,500 people, the first such loss in a decade.

People still remember oil drilling rigs lined up at the United States border, headed south in the early 1980s. They blame the NEP for the exodus of rigs, but they forget to mention what happened to the rigs when they arrived in Texas. They were cut up for scrap because there was no work for them there either!

The Cycle Goes On and On and On

During 2006 the price of oil rose to an all-time high again and then fell to more moderate levels. With no real shortage, no problems with the transportation system, people are scratching their heads. What was it all about this time? For more information read the newspapers for today's opinions, prognostications, and theories.

Supply and demand, transportation and technology, economics and politics, urbanization and fads, foreign policy, interest rates and debt, war and peace—all play a part. If the past is any help, we can make a few cautious predictions about the oil patch's future. The bust always ends when we least expect it, and for reasons we did not consider important. If the boom is short, the bust is often short too. If the boom lasts a long time, we often get sassy and expect it to last forever. When the bust comes, it can be very drastic.

We also should be careful not to take credit, or assign

blame, for the booms and busts. The Canadian oil industry is part of an international economic system that is far beyond the control of anyone in Canada. We can only control our reaction to the cycles. Spend like there's no tomorrow during a boom, if you like, but the bill eventually comes due.

Consumption patterns rise and fall in response to many forces, including weather, economic fortunes, and even wild-cards like wars. Fads can play a part too, though not for long. The bottom line is usually the price, because consumers think with their pocketbooks—and yet a month after a major price increase most people have forgotten about the shock.

Predicting the Boom and Bust Cycles

Governments sometimes influence the boom and bust cycles, but usually not according to their plan. If anything, government planning is consistent only in that it is usually the wrong reaction because, once the policies have made their way through the cumbersome political and bureaucratic system and start to take effect, the situation has already changed yet again. Every level of government usually follows the trends instead of leading them. Also, regional governments become increasingly antagonistic with each other during a boom and when the balance of fiscal power changes. Tension between parts of the country can become quite nasty. During the boom of the 1970s and early 1980s, for example, one bumper sticker read: "Let the eastern bastards freeze in the dark." Ralph Klein, the mayor of Calgary, blamed "eastern bums and creeps" for the rising crime rate. He said, "Don't go out and rob our banks and our convenience stores and mug our seniors and snatch purses—get the hell out of town!" On the continental level, it is important to remember that the American government never considers the needs of Canadian consumers, producers, or governments.

Governments can't predict the price of oil. Companies can't either.

To be fair, industry is not much better a predictor of the boom-bust cycles than government, nor is it more skilled at planning for the inevitable busts. In the early 1970s the Canadian oil patch said it had 923 years of oil and 392 years of natural gas reserves, but within a few years it was worried about supplies. Left to its own priorities—just making money—industry can be wasteful, pollute the environment, become lax on safety issues, and turn complacent. The mistakes and accidents that have happened in the history of the Canadian petroleum industry are many. Its arrogant attitude in the 1970s was "Get out of our way and let us produce the oil!" Its public face has improved over the decades, but the oil patch is still learning how to be a good citizen and how to balance its priorities with those of a modern society. When it comes to making plans for the downturn and its effects on society, industry usually relies on the government to take care of the needy.

Our society is becoming increasingly urbanized and will therefore use more and more hydrocarbons, still counting on an unlimited supply of energy. We want a reliable supply, and we want it cheap. As a result, with each boom and bust cycle our society becomes increasingly involved with the international economy and more susceptible to its twists and turns.

Surviving the Cycles

Booms and busts are like seasons in life. Each just lasts for a few years and, with a little creative thinking, you can enjoy the ride and perhaps even benefit from the rollercoaster.

During a boom you can …

Pace yourself so you can run the marathon and live through the boom.

Sell some real estate—housing prices rose 60 per cent in Calgary during 2005 and 2006.

Get a new job and save a quarter of your income or more if you can.

Avoid buying a new vehicle if possible.

Try to hang on to your friends and spend a little bit of time with them.

Travel overseas and rent out your home for a profit.

Maximize the economic upside to the boom, but don't get too caught up with the frantic energy of greed, busyness, and speculation.

Avoid making rash decisions, counting on the future being like the present, putting all your eggs in one basket, getting on the latest investment bandwagon, acting like things will stay like they are today, or taking advice from bankers, realtors, or salespeople.

Most of all, stop worrying—the boom will end soon.

"Oh me! Here we go—automobiles, colleges, taxes!" according to an undated cartoon in the *Oil and Gas Journal*. (Provincial Archives of Alberta, P1827)

During the bust you can …

Take time to breathe! Avoid new debt and other large expenses.

Buy a house or fix one up; these are long-term investments.

Get an education or add another degree to your resume.

Travel or take a longer holiday than usual; volunteer overseas.

Make new friends and get to know your neighbours better.

Reflect on the good things in your life, write an essay about your life, or sort pictures.

Take up a new hobby: make wine, learn crafts, take up cooking or gardening.

Have some kids and stay home with them for a few years.

Walk more for exercise and to reduce the need for a second vehicle.

Maximize the upside of the bust by taking life slower, doing things you did not have time for when you were busy; invest time and energy into parts of life that have less obvious economic value, like friendship and experiences. Take the long view.

Best of all, stop worrying—the bust will blow over soon.

"Blast! This wasn't here last Christmas," says Santa about the Calgary building boom in this Tom Innes cartoon that appeared December 24, 1981, in the *Calgary Herald*. (Glenbow Archives, M-8000-957)

Only the price at the pump
changes our habits—for a time.

Technology might affect the boom-bust cycle in small ways. A dramatically efficient car engine would create change for a few months or a year, until the price drops again. Electric cars and better public transportation can help too. Other fuel sources are promising, but still in development. But until the price at the pump rises so high that a litre of gasoline costs more than a litre of pop, we will probably just accept regular increases in the price of petroleum products without much concern.

Each boom and bust cycle is unique, different from those that came before. With each cycle we are further and further removed from the means of production. Resource extraction happens at the source, not in the city. In the early 1900s the oil-fields were near Calgary and Edmonton, but today's most productive fields are far away from urban centres—in Fort McMurray, the far North, and far offshore or overseas.

Finally, people have very short memories. During the busts we always look forward to the next boom. We remember the excitement of the good times, the increase in the value of our homes, the cheques from the provincial government, and the warm sense of well-deserved good fortune.

It's all warm and fuzzy until the rigs go to Texas and are cut up for scrap.

What You Can Do to Protect Yourself

You can make lots of decisions to help protect yourself from the worst parts of the boom and bust cycles.

First, don't trust short-term thinking. We all need work and a place to live, so find a job that you enjoy and a place to live you can afford regardless of current housing prices, interest rates, or the price of oil. If you decide to gamble on real estate, don't invest more than you are willing to lose.

Second, take advantage of today's conditions. During a boom you can make good money, pay off debts, and bank some, spend a bit more freely, and enjoy the good things that money can buy. When things get tight, sit tight. Stay closer to home and visit more with neighbours. Avoid accumulating debt, but if you can afford to buy a house, purchase during a downturn. If you want to change your career, go back to school during the recession when jobs are scarce and the cost of living is lower. Use some of the money you invested during the boom to improve your skills and get ready for the next boom.

Look out for yourself and others.

Third, have a plan, even if it's only sketchy. You may need to live in Calgary or Edmonton or Fort McMurray in order to make a living. But what about when you retire? Keep an eye out for an investment property that can one day become your retirement home, far from the booming city, if that's your dream, or buy a condo along the river close to your friends in the city.

Fourth, be ready for the unexpected. Though some people suffered during the downturn of the 1980s, others did fine. A house purchased by a young couple at the bottom of the market in the 1980s is now debt-free and holds hundreds of thousands of dollars in equity—maybe even more.

The oil patch is a gamble—that's what makes it so exciting. Rags to riches. Poverty to the penthouse. There's nothing wrong with gambling unless you expect to win every time.

Finally, find a way to be happy every day, or see the fun side.

As Bob Edwards, the irascible editor of the Calgary *Eye Opener* newspaper, noted, "The trouble with this oil situation at this formative stage is that you are never sure whether the man you meet on the street is a multi-millionaire, or just an ordinary, common millionaire." ∎

Glossary and Index

Terms in bold within the book are included below as both glossary items (with definitions), as well as index items (with corresponding page numbers). Other terms are included as index items only.

Abasand Oils Ltd., 103

acidizing – a way of improving the production from a well by forcing acid into the rock to open it up and allow the oil to come out. 8

air pollution, 92

Alaska Department of Environmental Conservation, 154

Alberta Energy Company, 69, 71, 115

Alberta Heritage Trust Fund – a fund created by Alberta premier Peter Lougheed's government in 1976 "to save for the future, to strengthen or diversify the economy, and to improve the quality of life of Albertans." During its first thirty years it generated $29 billion in investment income. 172, 173

Alberta Oil and Gas Conservation Board – today's Energy Resources Conservation Board. 105

Anderson, David, 153

Arctic Islands, 128

Arrow (oil tanker), 150

artificial lift – a method of extracting oil from the ground by pumping. 42

Athabasca River, 16, 18, 20, 100, 102, 113, 118, 119, 136, 139

Atlantic Accord – a 1985 agreement between the federal and Newfoundland and Labrador governments that provided for joint management of oil and gas resources off the East Coast of Canada. 60, 78

Atlantic No. 3 (oil well), 136, 145, 146

Atlantic Richfield, 107

atomic bomb, 103

aviation fuel – mid-grade oil used in jet engines. 12

bailer – a device for taking cuttings out of an oil well. 36

Bane, Ray, 159

barrel – 159 litres, 42 US gallons, or 35 Imperial gallons. 10, 45, 87

Beaufort Sea, 16, 120

Bell, Max, 103

Bell, Robert, 123

benchmark crude – a quality of oil against which others are compared. In North America the benchmark is West Texas Intermediate or WTI, with an API (American Petroleum Institute) gravity or weight of 39.6. A liquid with a gravity higher than 10 floats on top of water. WTI also contains about 0.24 per cent sulphur, making it a sweet—as opposed to sour— product. Other benchmark crudes include Brent Blend, used in Europe, and Dubai, used in the Middle East. 14

Bent Horn, Artic Islands, 120, 128

Berger Mackenzie Valley Pipeline Inquiry – a 1970s inquiry into a pipeline project in the Canadian Arctic. 10, 71, 130, 153

berm – pushed-up soil that acts as a wall. 147

biological productivity – a measure of the ability of an ecosystem to sustain life. 115

bits – drilling tools. 8

bitumen – heavy oil that is trapped in oil sands. 18, 98, 101, 109

Black Creek, Ontario, 139

Blair, Sid, 99

blowout – an oil well that goes out of control and literally blows out, spewing drilling mud, natural gas, oil, salt water, or other liquids out of the well. 8, 41, 160

blowout preventer (BOP) – equipment that stops a blowout. 2

booms and busts – economic cycles in the oil industry. 83, 163, 176, 178, 180, 181

Boreal Forest, 114

Bosworth Creek, Northwest Territories, 20

Bow Island field, Alberta, 20, 50

brackish water – salty water, but not as salty as sea water. 111

British American, 46

British Columbia, 150

British Commonwealth Air Training Plan, 168

British Petroleum, 93

Brown, Kootenay, 124, 125

butane – a component of natural gas. 4, 50

cable-tool rig – a type of drilling rig that uses a heavy bit with an up-and-down motion. 32, 34, 37

Calgary Petroleum Products, 18, 140

Calgary, Alberta, 50

Canadian American Norman Oil Line (CANOL) – a World War II pipeline through the Northwest and Yukon Territories. 10, 168

Canadian Association of Petroleum Producers (CAPP) – the representative body of the upstream oil and natural gas industry in Canada. CAPP represents 150 member companies that explore for, develop, and produce more than 95 per cent of Canada's natural gas, crude oil, oil sands, and elemental sulfur. 14

Canadian Energy Bank – a lender to oil companies. 72, 73

Canadian Petroleum Association (now called CAPP—see above), 69

Canadian Petroleum Safety Council, 41

Canadian Standards Association, 149

Canadian Western Natural Gas Company, 144

Canadianization – at least 50 per cent Canadian ownership of an industry. 76

Canmore, Alberta, 64

carbon, 110

Carney, Pat, 78, 79

carpooling, 94

casing – metal pipe that is put into an oil well to keep it from caving in. 8, 36

caustic soda – also called lye or sodium hydroxide, caustic soda is a strong chemical base often used in manufacturing applications. 111

Cherry Point, Washington, 151

Chevron, 93

Christmas tree – a valve assembly at the top of an oil well. 1

Chrystal, Lynn, 157

Cities Service, 105, 107

Clark, Joe, 61, 72

Clark, Karl, 98, 99

coal, 49, 63, 64, 122

coal bed methane (CBM) – natural gas in coal seams. 80, 120, 131

coal gas, 164

coal oil – low-quality fuel. 164

cogeneration stations – facilities that use natural gas to make electricity and heat. 54

Cohasset, Atlantic Ocean, 120, 128

cold production – a method for extracting oil from the oil sands without heat. 111

Commons Committee on Environmental Pollution, 153

commuting, 52, 57, 95

Competition Bureau, 88

compression – pressurizing of natural gas. 55

computor – a person who did computations on an early seismic crew. 6

condensate – liquids in natural gas. 1

Confederation, 63

conventional oil – crude oil found in the ground. 58

core drilling – a method of taking a sample of rock. 103

core sample, 25

Coste, Eugene, 24

Cottrelle, George (Dominion Oil Controller), 66

Covery, Cliff, 146

Crowsnest Pass, Alberta, 64

crude oil – unrefined petroleum. 4

crude oil prices, 89

cyclic steam stimulation – a method of melting oil sand underground. 111

Davidson, Art, 154

derrick – a drilling rig. 6

derrickhand or derrickman – a person who works near the top of the rig. 4, 42

Devon, Alberta, 127

Devonian-aged reefs, 126, 127

Dick, Dr. D., 103

Diefenbaker, John, 51, 68

diesel – one product of crude oil. 12, 48

diluents – materials that dilute heavy oil. 110

directional drilling – angled drilling, usually on purpose. 8, 41

doghouses – shacks on drilling rigs. 4

doodlebugs – seismic workers. 6

Dover field, New Brunswick, 19

downstream – the part of the industry that delivers oil to the consumer. 13

dowsers – people who finds water and oil with forked sticks. 6, 28

drill ships and platforms – boats and floating platforms that serve as a temporary ocean-going location for a drilling rig as it punctures a hole into the surface of the earth beneath a body of water. 8

driller – the person who drills an oil well. 4, 24, 42

drilling contractors, 39

drilling log – a record of drilling activity. 41

drilling platforms, 8

drilling rig – a machine that drills oil wells. 2, 6

dry hole – a skunked or failed oil well. 7

Edmonton, Alberta, 50

electric cars, 180

electrical logging – a method of testing, or logging, the rock formations in an oil well. 28

elephants – large discoveries of oil or gas. 166

EnCana Corporation, 131

energy economists, 166

Energy Resources Conservation Board (ERCB) – a board created in 1938 to oversee the development of the Alberta oil industry and control economic waste (also called the Petroleum and Natural Gas Conservation Board as well as the Alberta Energy and Utilities Board). 115

Energy Self-Sufficiency Tax, 60, 72, 73

Enform, 41

engineers, 53

Enniskillin, Ontario, 19

Environmental Impact Assessment – a study of the effects of petroleum exploration and extraction on the environment. 115

ethane – a chemical compound that is part of natural gas and a by-product of the refining of petroleum. 4

Evans, Cal, 122

explorationists – geologists or geophysicists. 5

Exxon Valdez (oil tanker), 136, 155

Exxon, 158

ExxonMobil, 93

faulted – cracked rock. 9

feedstock – a source material for other products. 53

field lines – a system of pipelines for collecting oil in the field. 10

fireflooding – a method by which oxygen and fire help melt oil sands underground. 111

fish – to pull lost equipment out of a well. 36

Fitzsimmons, Robert, 102

flare – a tower that burns off natural gas. 5

floorhands, 42

Ford, Henry, 47

Foremost (equipment), 29

formation – a place where oil is stored in rock. 2

Fort Fitzgerald, Alberta, 118

Fort McMurray, Alberta, 19, 101, 108, 116

Fort Smith, Alberta, 118

fractionated – a formation that has been opened with pressure. 12

fractured – cracked, referring to rock formations. 9

frontiers – the edges of the landscape, physical or mental. 8, 30, 31, 80, 121, 123, 126, 129, 134

fuel oil – a type of oil used in furnaces. 13, 48, 87

gas hydrates – natural gas trapped in crystals. 120, 134, 167

gas plant – facility that processes natural gas. 12, 53, 55, 170

gasoline, 12, 13, 47, 50, 55, 56, 85, 86, 87, 88, 93, 95, 114

Geological Survey of Canada (GSC) – part of the Earth Sciences Sector of Natural Resources Canada, the GSC performs geological surveys of the country to help develop the country's natural resources. 23, 62, 66, 98, 123

geologists – scientists who study the physical processes of the earth in order to understand its history. 4, 5, 23, 24, 126

geophones – listening devices used to capture sound waves by geophysicists who then use the information to understand the earth. 29

geophysicist – a scientist who studies the physical properties of the earth. 6, 23, 29

geophysics, 28

Gesner, Abraham, 19, 164

Gleichen, Alberta, 28

Good Roads Movement, 47

Gow, John A., 163

Grand Banks, Newfoundland, 128

grandfather – to abide by a previous agreement after new laws or rules have come into effect. 79

Grant, George, 122

Gray, Jim, 69, 75

Great Canadian Oil Sands (Suncor), 98, 105

Great Slave Lake, Northwest Territories, 118

Green, Howard, 104

greenhouse gases – gases that contribute to global warming: carbon dioxide, methane, nitrous oxide, and water vapour. 110, 115

Gulf Oil Canada, 71, 107, 153

gum beds, 33

gusher – an oil well that is spurting oil or gas. 8, 34, 163

gypsum – a soft mineral made up of calcium sulfate dihydrate, most commonly used to make wallboard for the interior walls in houses. 111

Harper, Stephen, 81

heating oil – oil used in furnaces. 12, 55

heavy oil – thick, molasses-like petroleum. 55, 82, 98, 99

Hell's Half Acre – coulee north of the village of Turner Valley where oil companies once burned off waste gas while producing oil. 141

Herron, Bill, 18, 140

Hibernia (drilling platform), 16, 120, 128

Hoffman, G. C., 123

horizontal drilling – drilling parallel to the earth's surface. 8

House of Commons Standing Committee on National Resources and Public Works, 122

Hume, Dr. G. S., 66

Hunt, Thomas Sterry, 22

hydrogen sulphide – a deadly component of natural gas. 5, 53, 64, 142, 170

Iarossi, Frank, 156

Imperial Oil, 38, 46, 65, 71, 107

Imperial-Leduc No. 1 (oil well), 26

impurities – salt, water, sulphur, or other pollutants found in oil or gas. 5, 9, 12, 53

in situ – in site. 54, 112

incentives, 79

internal combustion – explosions in a confined place, such as a motor. 47, 165

International Bitumen Company, 102

international commodity speculation – a way of trading resources. 14

Interprovincial Pipe Line Company, 147

Iran-Iraq war, 174

Irving Whale (oil barge), 150

Jones, B. O., 103

Kananaskis, Alberta, 173

Kelsey, Henry, 18, 100, 120, 123

keroselain – the original name for kerosene, derived from "wax oil" in the Greek. 164

kerosene – mid-grade oil. 12, 19, 46, 164

King, William Lyon Mackenzie, 66

Kipling, Rudyard, 124

Klein, Ralph, 77, 81, 176

Knode, Bill, 143

Lake Erie, Ontario, 139

Lalonde, Marc, 61

landmen – people, male or female, who negotiate with land or mineral rights owners for access to a property for the purpose of exploring for oil and gas or for production of minerals. 30

Langevin, Alberta, 16, 20, 120, 124

Laurier, Wilfrid, 66

Leduc, Alberta, 16, 126, 146, 165

Leduc-Woodbend field, Alberta, 26

Leiper, Hugh, 145

Link, Ted, 20

liquefied natural gas (LNG) – natural gas in a liquid state, as opposed to vapour. 11

LNG tanker – a ship that moves liquid natural gas. 11

Lodgepole, Alberta, 160

Lougheed, Peter, 69, 73, 82, 106, 172

low-pressure pipeline – a pipeline that delivers natural gas to a residence. 13

Macdonald, Sir John A., 63

Mackenzie Delta, Northwest Territories, 71, 120, 128, 134, 153

Mackenzie River, 16, 20, 30, 101, 118, 125, 129

Mackenzie Valley Pipeline – an oil and gas transmission line in the North. 10, 71, 153

Mackenzie, Alexander, 101, 123

Macoun, John, 119, 123

Malik, Northwest Territories, 134

Manning, Ernest, 106

Mansfields Water and Oil Finder, 28

map – a method that records hints about the location of petroleum. 6, 23, 27, 28, 29, 30

mature fields – older oilfields that are nearing the end of their production. 8, 80

MCF (or thousand cubic feet) – the standard unit of measure for natural gas. 13, 170

McMahon, Frank, 146

Medicine Hat, Alberta, 20

mercaptors – substances added to natural gas to give it an unpleasant smell. 12

methane – a component of natural gas. 4, 58, 134

midstream – the part of the industry that refines the oil. 5, 10, 12

Miller, George, 156

mineral rights – the permission to exploit a resource. 30, 131

Minister of National Finance, 67

Mississippian, 127

Mitchell, W. O., 142

Moberly, Henry, 123

Mobil Oil Canada, 71, 153

monkeyboard – a platform high in a rig. 4

motorhand, 42

mud – sloppy mixture used to circulate in an oil well and remove cuttings. 36, 39

Mulroney, Brian, 78

naphtha – a liquid similar to gasoline, also called casing-head gasoline. 5, 141, 143

National Energy Board (NEB) – a national petroleum industry regulator. 113, 115, 119, 148, 169

National Energy Program (NEP) – a set of energy policies implemented by the government of Canada in 1980 with the intent of promoting self-sufficiency, Canadian ownership of the industry, and security of supply. 61, 74, 75, 77, 81, 83, 172

National Oil Policy – Ottawa's 1960 oil policy. 60, 68

National Policy – Ottawa's 1870 economic policy. 60, 62, 63, 64

natural gas – the vaporous part of oil. 4, 50, 114, 139, 140, 142, 144, 167, 170, 171, 174, 177

Natural Resources Revenue Plan – an Alberta plan to increase oil and gas royalties instituted in 1973 by the Lougheed government. 60, 70

New Brunswick, 127

Newfoundland and Labrador, 78, 128, 160

NGLs – natural gas liquids. 5

nitroglycerine – an explosive used to open up geological formations. 37

Nixon, Richard, 68

Nodwell, 29

Norman Wells, Northwest Territories, 120, 169

Norman Wells Pipeline, 16

North American Free Trade Agreement (NAFTA) – a trade agreement signed by the United States, Canada, and Mexico in 1994. 81

North Saskatchewan River, 147

Northwest Territories, 80

Nova Scotia, 16, 150

Ocean Ranger (drilling platform), 136, 160

offshore drilling – rigs or platforms that allow drilling at sea. 8, 42, 128, 167

oil and gas resources, 121

Oil City, Alberta, 125

oil sands – sands that contain bitumen, a tar-like oil, that are found in Alberta, the western United States, and Venezuela. 54, 58, 80, 81, 82, 98, 99, 100, 105, 106, 107, 108, 109, 110, 113, 114, 115, 116, 118, 131, 167

oil shales – fine-grained sedimentary rock that contains hydrocarbons. 19, 134

Oil Springs, Ontario, 33, 34, 120, 126

oil stoves, 47

oil well, 3, 6, 8, 16, 18, 21, 33, 42, 126

oilfield – area of land where oil and gas are found underground. 3, 53

Organization of Petroleum Exporting Countries (OPEC) – a group of oil-producing countries that joined together in 1960 in order to sell their products at a fair price. OPEC members include Algeria, Angola, Indonesia, Iran, Iraq, Kuwait, Libya, Nigeria, Qatar, Saudi Arabia, the United Arab Emirates, and Venezuela. 72, 89, 172

orphan wells – wells whose owners cannot be found. 2

Ottawa Valley line, 68

overburden – soil removed to get at oil sands. 114

Panarctic Oils, 71, 128

Pelican Point, Alberta, 20

Pelican Rapids, Alberta, 139

pentane – a component of natural gas. 4, 50

perforating – popping holes in the side of a well to increase flow. 8

Petro-Canada, 71, 75, 88

petroleum accountants, 166

Petroleum and Natural Gas Conservation Board – a predecessor to the Energy Resources Conservation Board that regulated the development of the oil and gas industry in Alberta, beginning in 1938. 143, 144

Petroleum Bounty Act – 1904 federal legislation that provided a subsidy for the production of oil. 60, 66

Petroleum Industry Training Service, 41

Petrolia, Ontario, 16

Pew, J. Howard, 105

pigs – devices that clean out pipelines from the inside. 4

pipeline workers, 56

pipelines, 56, 68, 126

plays – areas or regions that are developed for their oil and gas potential. 9

Pond, Peter, 18, 100, 123

pool – place where oil is trapped in rock. 2, 4

price of crude oil – the international price of oil. 14, 86, 88, 122

Prince Edward Island, 128, 136, 150

producer – a productive oil well. 7

production – a successful oil well. 8, 58

productivity, 9

profits, 93, 96

Project Cauldron – a 1958 plan to use a nuclear bomb to melt the Alberta oil sands deep underground. 103

propane – a component of natural gas. 4, 50

proppant – sand or other materials used to prop open cracks in an oil well. 2

prospect – an area of exploration. 77, 153

pump jack – a pump that sucks oil out of an oil well. 3, 42

quad – one quadrillion BTUs (1,055 petajoules), a trillion cubic feet of natural gas. 1

Rawl, Lawrence, 159

Rebus, John, 146

reclamation – the process of returning mined land to useful status. 115

records – ways of saving information from a seismic survey. 6, 29, 30

Redwater field, Alberta, 26

refined oil – oil that is processed through a refinery. 4

refinery – a plant that makes lubricant, oil, and gasoline from crude oil. 4, 12, 46, 47, 88, 110, 169

Research Council of Alberta, 99

reserves – amounts of oil and gas in the earth. 4, 58, 71, 73, 80, 122, 165, 177

reservoir – a place where oil is trapped in rock. 7, 53

residential gas company – supplier of natural gas to consumers. 13

restoration – the process of returning mined land to useful status. 114

Richfield Oil Company, 103

rig manager, 42

Ross, Victor, 19

rotary drilling rigs – drilling rigs that use a rotating drill bit. 32, 37, 39

roughneck – a person who works on a drilling rig. 2, 42

Rowland, Les, 147

Royalite No. 4 (oil well), 140

royalty – the rent paid for taking a resource. 70, 79, 82, 172

rule of capture – the legalized theft of hydrocarbons. 145

runaway – an oil well that is out of control. 8

Sable Island, Atlantic Ocean, 120, 128

salt domes – underground storage for oil. 4

Santa Barbara, California, 150

Sarnia, Ontario, 47

Schlumberger, Conrad and Marcel, 28

Scott, Jim, 159

secondary recovery – methods of making wells productive. 8

Seepage Creek, Alberta, 139

seismic – sound waves used to look for oil. 6, 23, 28, 30

self-sufficient, 49, 73

service rig – equipment that repairs an oil well. 8

service station, 46, 47

settling ponds – dug-out holes in the ground where contaminated water stands in order that suspended particles can settle to the bottom. 110

Seven Sisters, 93

Shaw, Hugh Nixon, 33, 138

Shell Canada, 105

Shell, 93

shot hole – a hole where an explosive is detonated to make a seismic signal. 29

Slave River, 118

slurry – a mixture of liquids and solids, such as water and clay. 54

SmartCar, 57

smog – the haze formed by burning fossil fuels. 89, 92

sour gas – natural gas that contains hydrogen sulphide. 5, 64, 143

source rock – the rock formation where oil is created and stored. 2

South – slang for accident or blowout. 41

spill (oil), 145, 147, 150, 151, 154, 155, 159

spring-pole drilling system – drilling using a bouncy log. 34

Stalnaker, Charles, 33

Standard Oil, 65

steam assisted gravity drainage (SAGD) – an underground oil sand recovery method. 111

Stoney Creek field, New Brunswick, 19

Suez Canal, 105

Suez Crisis, 49, 168

sulphur – a naturally occurring element found in petroleum. 5, 50, 64

Sun Oil Co., 103, 104

Suncor, 106

surface rights – a legal entitlement of ownership of the topmost portion of an area of land, as opposed to subsurface or mineral rights. 30

surveys – a method of looking for oil and gas or hints of it. 5

sweet gas – sour gas that has been cleaned of impurities. 5

sweet water – potable or drinkable water. 9

Syncrude, 106, 107

synthetic oil – created by upgrading heavy oil from the oil sands. 110

Taglu gas discovery, Northwest Territories, 120, 128

tailings ponds – dug-out holes in the ground where contaminated water stands. 110, 113

Tanner, N. E., 103

tar sands – now usually called oil sands, tar sands are deposits of sand that contain bitumen, a thick petroleum product the consistency of molasses. 99

taxes, 72, 78, 79, 82, 88

toe-to-heel air injection (THAI) – a method by which oxygen and fire encourage melting of oil sands underground. 111

three-dimensional seismic, or 3D – a process that is done in three dimensions; for example, 3D seismic records information on the surface as well as down into the earth in two more planes. 30

tools – bits and other drilling equipment. 8

Trans Canada Airlines (now called Air Canada), 149

Trans-Alaska Pipeline System (TAPS), 151

TransCanada Pipelines Limited, 27, 149

transmission – moving oil or gas through a pipeline. 10

transportation needs, 94, 95

traps – places where oil is stored in rock, like pools. 4

trip out – to take out all pipe, add a new bit, and run the pipe back into the well. 41

Trudeau, Pierre, 61, 71, 74, 106

Turner Valley, Alberta, 16, 20, 25, 39, 120, 136, 140, 142, 165, 170

Turner Valley Royalties No. 1 (oil well), 142

turpentine – a product of oil, used for thinning oil-based paints, 12

Tweedle, H. C., 19, 127

Ultramar, 88

upgrader – a refinery that increases the value of an oil product. 110

upstream – the part of the industry that finds and drills for oil. 5, 6, 41

Valdez, Alaska, 151, 157

Vancouver, British Columbia, 47

vapour recovery extraction (VAPEX) – an underground oil sand recovery method. 111

vibroseis – a method used in seismic processing whereby signals are generated through the earth using vibrations. 15

Viking field, Alberta, 50

VOCs – volatile organic compounds, gases and vapours. 2

Wa-Pa-Su, 18, 120

Wartime Oils Limited – a government program that stimulated oil exploration during World War II. 60, 66, 67

waste gas – natural gas that is flared. 5

water, 110, 112, 113

water-flooding – a way of adding pressure to an oilfield. 166

Waterton, Alberta, 16, 20, 120, 125, 139

well-log analysts – people who interpret information from well testing, or logging, in order to understand the rock formations in an oil well. 53

West Texas Intermediate price – the international base price for oil used in North America. 14

Western Accord – the 1985 agreement between Ottawa and the western provinces over the price of oil. 60, 78

Western Canada Sedimentary Basin – where oil is found in the West under portions of southwestern Manitoba, southern Saskatchewan, most of Alberta, northeastern British Columbia, and the southwestern part of the Northwest Territories.

Whitehorse, Yukon, 169

wildcat wells – wells drilled in a new area, exploratory ventures. 7

Will, Ralph, 39

Williams, J. H., 33

Winchell, Professor Alexander, 138

windfall profits – unexpected profits. 83

wooden derricks, 34

World Wars I and II, 168

write-offs – allowances made by a government for expenses. 108